T0076753

Supervised Learning with Python

Concepts and Practical Implementation Using Python

Vaibhav Verdhan

Foreword by Dr. Eli Yechezkiel Kling (PhD)

Apress®

Supervised Learning with Python: Concepts and Practical Implementation Using Python

Vaibhav Verdhan
Limerick, Ireland

ISBN-13 (pbk): 978-1-4842-6155-2 ISBN-13 (electronic): 978-1-4842-6156-9
https://doi.org/10.1007/978-1-4842-6156-9

Managing Director, Apress Media LLC: Welmoed Spahr
Acquisitions Editor: Celestin Suresh John
Development Editor: James Markham
Coordinating Editor: Shrikant Vishwakarma

Cover designed by eStudioCalamar

Cover image designed by Pexels

Distributed to the book trade worldwide by Springer Science+Business Media New York, 233 Spring Street, 6th Floor, New York, NY 10013. Phone 1-800-SPRINGER, fax (201) 348-4505, e-mail orders-ny@springer-sbm.com, or visit www.springeronline.com. Apress Media, LLC is a California LLC and the sole member (owner) is Springer Science + Business Media Finance Inc (SSBM Finance Inc). SSBM Finance Inc is a **Delaware** corporation.

For information on translations, please e-mail booktranslations@springernature.com; for reprint, paperback, or audio rights, please e-mail bookpermissions@springernature.com.

Apress titles may be purchased in bulk for academic, corporate, or promotional use. eBook versions and licenses are also available for most titles. For more information, reference our Print and eBook Bulk Sales web page at http://www.apress.com/bulk-sales.

Any source code or other supplementary material referenced by the author in this book is available to readers on GitHub via the book's product page, located at www.apress.com/978-1-4842-6155-2. For more detailed information, please visit http://www.apress.com/source-code.

Printed on acid-free paper

To Yashi, Pakhi and Rudra.

Table of Contents

About the Author

Vaibhav Verdhan has 12+ years of experience in data science, machine learning, and artificial intelligence. An MBA with engineering background, he is a hands-on technical expert with acumen to assimilate and analyze data. He has led multiple engagements in ML and AI across geographies and across retail, telecom, manufacturing, energy, and utilities domains. Currently he resides in Ireland with his family and is working as a Principal Data Scientist.

About the Technical Reviewer

Irfan Elahi is a full stack customer-focused cloud analytics specialist bearing the unique and proven combination of diverse consulting and technical competencies (cloud, big data, and machine learning) with a growing portfolio of successful projects delivering substantial impact and value in multiple capacities across telecom, retail, energy, and health-care sectors. Additionally, he is an analytics evangelist as is evident from the published book, Udemy courses, blogposts, trainings, lectures, and presentations with global reach.

Foreword

How safe is home birthing? That is a good question. Pause a moment and let yourself contemplate it.

I am sure you can see how the answer to this question can affect personal decisions and policy choices. The answer could be given as a probability, a level classification, or an alternative cost. Another natural reaction is "it depends." There are many factors that could affect the safety of home birthing.

I took you through this thought exercise to show you that you naturally think like a data scientist. You understood the importance of stipulating clearly the focus of the analysis and what could explain different outcomes. The reason you are embarking on a journey through this book is that you are not sure how to express these instinctive notions mathematically and instruct a computer to "find" the relationship between the "Features" and the "Target."

When I started my career 30-odd years ago, this was the domain of statisticians who crafted a mathematical language to describe relationships and noise. The purpose of predictive modeling was in its essence to be a tool for separating a signal or a pattern out of seemingly chaotic information and reporting how well the partition was done.

Today, machine learning algorithms harnessing computing brute force add a new paradigm. This has created a new profession: the data scientist. The data scientist is a practitioner who can think in terms of statistical methodology, instruct a computer to carry out the required processing, and interpret the results and reports.

Becoming a good data scientist is a journey that starts with learning the basics and mechanics. Once you are done exploring this book you might also be able to better see where you will want to deepen your theoretical knowledge. I would like to suggest you might find it interesting to look into the theory of statistical modeling in general and the Bayesian paradigm specifically. Machine learning is computational statistics after all.

Dr. Eli. Y. Kling (BSc. Eng. MSc. PHD) London, UK. June 2020.

Acknowledgments

I would like to thank Apress publications, Celestin John, Shrikant Vishwarkarma, and Irfan Elahi for the confidence shown and the support extended. Many thanks to Dr. Eli Kling for the fantastic forward to the book. Special words for my family—Yashi, Pakhi, and Rudra—without their support it would have been impossible to complete this work.

Introduction

> *"It is tough to make predictions, especially about the future."*
>
> —Yogi Berra

In 2019, MIT's Katie Bouman processed five petabytes of data to develop the first-ever image of a black hole. Data science, machine learning, and artificial intelligence played a central role in this extraordinary discovery.

Data is the new electricity, and as per *HBR*, data scientist is the "sexiest" job of the 21st century. Data is fueling business decisions and making its impact felt across all sectors and walks of life. It is allowing us to create intelligent products, improvise marketing strategies, innovate business strategies, enhance safety mechanisms, arrest fraud, reduce environmental pollution, and create path-breaking medicines. Our everyday life is enriched and our social media interactions are more organized. It is allowing us to reduce costs, increase profits, and optimize operations. It offers a fantastic growth and career path ahead, but there is a dearth of talent in the field.

This book attempts to educate the reader in a branch of machine learning called *supervised learning*. This book covers a spectrum of supervised learning algorithms and respective Python implementations. Throughout the book, we are discussing building blocks of algorithms, their nuts and bolts, mathematical foundations, and background process. The learning is complemented by developing actual Python code from scratch with step-by-step explanation of the code.

The book starts with an introduction to machine learning where machine learning concepts, the difference between supervised, semi-supervised, and unsupervised learning approaches, and practical use cases are discussed. In the next chapter, we examine regression algorithms like linear regression, multinomial regression, decision tree, random forest, and so on. It is then followed by a chapter on classification algorithms using logistic regression, naïve Bayes, knn, decision tree, and random forest. In the next chapter, advanced concepts of GBM, SVM, and neural network are studied. We are working on structured data as well as text and image data in the book. Pragmatic Python implementation complements the understanding. It is then followed by the final chapter on end-to-end model development. The reader gets Python code, datasets, best practices, resolution of common issues and pitfalls, and pragmatic first-hand knowledge on implementing algorithms. The reader will be able to run the codes and extend them in an innovative manner, as well as will understand how to approach a supervised learning problem. Your prowess as a data science enthusiast is going to get a big boost, so get ready for these fruitful lessons!

The book is suitable for researchers and students who want to explore supervised learning concepts with Python implementation. It is recommended for working professionals who yearn to stay on the edge of technology, clarify advanced concepts, and get best practices and solutions to common challenges. It is intended for business leaders who wish to gain first-hand knowledge and develop confidence while they communicate with their teams and clientele. Above all, it is meant for a curious person who is trying to explore how supervised learning algorithms work and who would like to try Python.

Stay blessed, stay healthy!

—Vaibhav Verdhan
Limerick,
Ireland. June 2020

CHAPTER 1

Introduction to Supervised Learning

> *"The future belongs to those who prepare for it today."*
>
> — Malcom X

The future is something which always interests us. We want to know what lies ahead and then we can plan for it. We can mold our business strategies, minimize our losses, and increase our profits if we can predict the future. Predicting is traditionally intriguing for us. And you have just taken the first step to learning about predicting the future. Congratulations and welcome to this exciting journey!

You may have heard that data is the new oil. Data science and machine learning (ML) are harnessing this power of data to generate predictions for us. These capabilities allow us to examine trends and anomalies, gather actionable insights, and provide direction to our business decisions. This book assists in developing these capabilities. We are going to study the concepts of ML and develop pragmatic code using Python. You are going to use multiple datasets, generate insights from data, and create predictive models using Python.

By the time you finish this book, you will be well versed in the concepts of data science and ML with a focus on supervised learning. We will examine concepts of supervised learning algorithms to solve regression

V. Verdhan, *Supervised Learning with Python*,
https://doi.org/10.1007/978-1-4842-6156-9_1

problems, study classification problems, and solve different real-life case studies. We will also study advanced supervised learning algorithms and deep learning concepts. The datasets are structured as well as text and images. End-to-end model development and deployment process are studied to complete the entire learning.

In this process, we will be examining supervised learning algorithms, the nuts and bolts of them, statistical and mathematical equations and the process, what happens in the background, and how we use data to create the solutions. All the codes use Python and datasets are uploaded to a GitHub repository (`https://github.com/Apress/supervised-learning-w-python`) for easy access. You are advised to replicate those codes yourself.

Let's start this learning journey.

What Is ML?

When we post a picture on Facebook or shop at Amazon, tweet or watch videos on YouTube, each of these platforms is collecting data for us. At each of these interactions, we are leaving behind our digital footprints. These data points generated are collected and analyzed, and ML allows these giants to make logical recommendations to us. Based on the genre of videos we like, Netflix/YouTube can update our playlist, what links we can click, and status we can react to; Facebook can recommend posts to us, observing what type of product we frequently purchase; and Amazon can suggest our next purchase as per our pocket size! Amazing, right?

The short definition for ML is as follows: "In Machine Learning, we study statistical/mathematical algorithms to learn the patterns from the data which are then used to make predictions for the future."

And ML is not limited to the online mediums alone. Its power has been extended to multiple domains, geographies, and use cases. We will be describing those use cases in detail in the last section of this chapter.

So, in ML, we analyze vast amounts of data and uncover the patterns in it. These patterns are then applied on real-world data to make predictions for the future. This real-world data is unseen, and the predictions will help businesses shape their respective strategies. We do not need to explicitly program computers to do these tasks; rather, the algorithms take the decisions based on historical data and statistical models.

But how does ML fit into the larger data analysis landscape? Often, we encounter terms like data analysis, data mining, ML, and artificial intelligence (AI). Data science is also a loosely used phrase with no exact definition available. It will be a good idea if these terms are explored now.

Relationship Between Data Analysis, Data Mining, ML, and AI

Data mining is a buzzword nowadays. It is used to describe the process of collecting data from large datasets, databases, and data lakes, extracting information and patterns from that data, and transforming these insights into usable structure. It involves data management, preprocessing, visualizations, and so on. But it is most often the very first step in any data analysis project.

The process of examining the data is termed *data analysis*. Generally, we trend the data, identify the anomalies, and generate insights using tables, plots, histograms, crosstabs, and so on. Data analysis is one of the most important steps and is very powerful since the intelligence generated is easy to comprehend, relatable, and straightforward. Often, we use Microsoft Excel, SQL for EDA. It also serves as an important step before creating an ML model.

There is a question quite often discussed—what is the relationship between ML, AI, and deep learning? And how does data science fit in? Figure 1-1 depicts the intersections between these fields. AI can be thought of as automated solutions which replace human-intensive tasks. AI hence reduces the cost and time consumed as well as improving the overall efficiency.

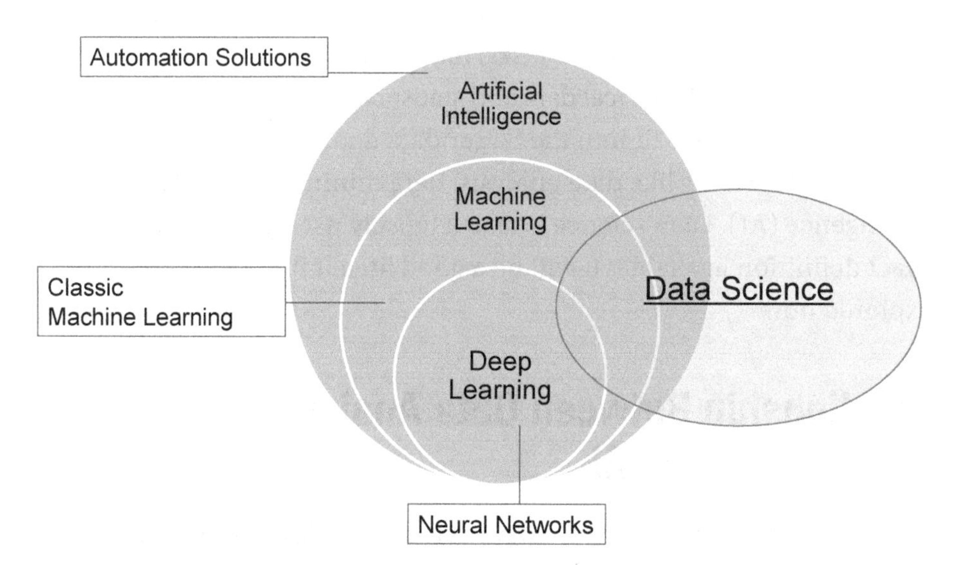

Figure 1-1. *Relationship between AI, ML, deep learning, and data science shows how these fields are interrelated with each other and empower each other*

Deep learning is one of the hottest trends now. Neural networks are the heart and soul of deep learning. Deep learning is a subset of AI and ML and involves developing complex mathematical models to solve business problems. Mostly we use neural networks to classify images and analyze text audio and video data.

Data science lies at the juxtaposition of these various domains. It involves not only ML but also statistics understanding, coding expertise and business acumen to solve business problems. A data scientist's job is to solve business problems and generate actionable insights for the business. Refer to Table 1-1 to understand the capabilities of data science and its limitations.

Table 1-1. *Data Science: How Can It Help Us, Its Usages, and Limitations*

How data science can help	Limitations of data science
Assist in making decisions by analyzing multi dimensional data which is quite difficult for a human being	Data is not an alternative to experience
Use statistical tools & techniques to uncover patterns	Data science cannot replace the subject matter knowledge
The algorithms further help in measuring the accuracy of the patterns & the claims	Data science depends on data availability and data quality. Depending on the input,we will get the output
The results are reproducible and can be improved	Data science will not increase the revenue or sales or output by 50% overnight. Similarly, it will not decrease the cost by 1/3 immediately
The machine learns, which is a big difference from the traditional software engineering	A data science project takes time to be implemented

With the preceding discussion, the role of ML and its relationship with other data-related fields should be clear to you. You would have realized by now that “data” plays a pivotal role in ML. Let’s explore more about data, its types and attributes.

Data, Data Types, and Data Sources

You already have some understanding of data for sure. It will be a good idea to refresh that knowledge and discuss different types of datasets generated and examples of it. Figure 1-2 illustrates the differentiation of data.

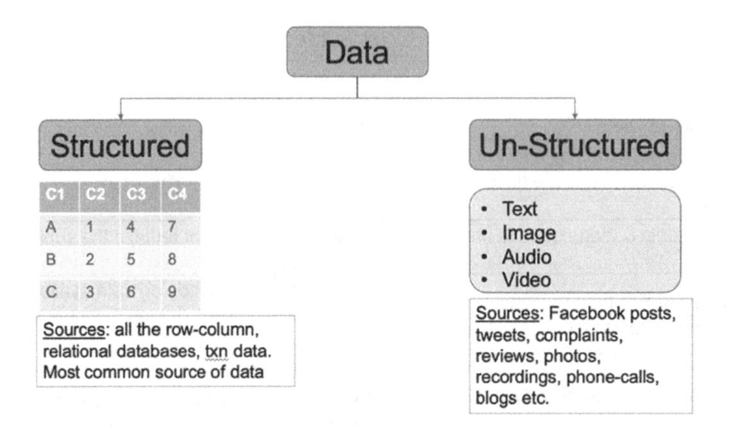

Figure 1-2. *Data can be divided between structured and unstructured. Structured data is easier to work upon while generally deep learning is used for unstructured data*

Data is generated in all the interactions and transactions we do. Online or offline: we generate data every day, every minute. At a bank, a retail outlet, on social media, making a mobile call: every interaction generates data.

Data comes in two flavors: structured data and unstructured data. When you make that mobile call to your friend, the telecom operator gets the data of the call like call duration, call cost, time of day, and so on. Similarly, when you make an online transaction using your bank portal, data is generated around the amount of transaction, recipient, reason of transaction, date/time, and so on. All such data points which can be represented in a row-column structure are called *structured data*. Most of the data used and analyzed is structured. That data is stored in databases and servers using Oracle, SQL, AWS, MySQL, and so on.

Unstructured data is the type which cannot be represented in a row-column structure, at least in its basic format. Examples of unstructured data are text data (Facebook posts, tweets, reviews, comments, etc.), images and photos (Instagram, product photos), audio files (jingles, recordings, call center calls), and videos (advertisements, YouTube posts, etc.). All of the unstructured data can be saved and analyzed though. As you would imagine, it is more difficult to analyze unstructured data than structured data. An important point to be noted is that unstructured data too has to be converted into integers so that the computers can understand it and can work on it. For example, a colored image has pixels and each pixel has RGB (red, green, blue) values ranging from 0 to 255. This means that each image can be represented in the form of matrices having integers. And hence that data can be fed to the computer for further analysis.

Note We use techniques like natural language processing, image analysis, and neural networks like convolutional neural networks, recurrent neural networks, and so on to analyze text and image data.

A vital aspect often ignored and less discussed is *data quality*. Data quality determines the quality of the analysis and insights generated. Remember, *garbage in, garbage out.*

The attributes of a good dataset are represented in Figure 1-3. While you are approaching a problem, it is imperative that you spend a considerable amount of time ascertaining that your data is of the highest quality.

Figure 1-3. *Data quality plays a vital role in development of an ML solution; a lot of time and effort are invested in improving data quality*

We should ensure that data available to us conforms to the following standards:

- **Completeness** of data refers to the percentage of available attributes. In real-world business, we find that many attributes are missing, or have NULL or NA values. It is advisable to ensure we source the data properly and ensure its completeness. During the data preparation phase, we treat these variables and replace them or drop them as per the requirements. For example, if you are working on retail transaction data, we have to ensure that revenue is available for all or almost all of the months.

- **Data validity** is to ensure that all the key performance indicators (KPI) are captured during the data identification phase. The inputs from the business subject matter experts (SMEs) play a vital role in ensuring this. These KPIs are calculated and are verified by the SMEs. For example, while calculating the average call cost of a mobile subscriber, the SME might suggest adding/deleting few costs like spectrum cost, acquisition cost, and so on.
- **Accuracy** of the data is to make sure all the data points captured are correct and no inconsistent information is in our data. It is observed that due to human error or software issues, sometimes wrong information is captured. For example, while capturing the number of customers purchasing in a retail store, weekend figures are mostly higher than weekdays. This is to be ensured during the exploratory phase.
- Data used has to be **consistent** and should not vary between systems and interfaces. Often, different systems are used to represent a KPI. For example, the number of clicks on a website page might be recorded in different ways. The consistency in this KPI will ensure that correct analysis is done, and consistent insights are generated.
- While you are saving the data in databases and tables, often the relationships between various entities and attributes are not consistent or worse may not exist. Data **integrity** of the system ensures that we do not face such issues. A robust data structure is required for an efficient, complete, and correct data mining process.

- The goal of data analytics is to find trends and patterns in the data. There are seasonal variations, movements with respect to days/time and events, and so on. Sometimes it is imperative that we capture data of the last few years to measure the movement of KPIs. The **timeliness** of the data captured has to be representative enough to capture such variations.

Most common issues encountered in data are missing values, duplicates, junk values, outliers, and so on. You will study in detail how to resolve these issues in a logical and mathematical manner.

By now, you have understood what ML is and what the attributes of good-quality data are to ensure good analysis. But still a question is unanswered. When we have software engineering available to us, why do we still need ML? You will find the answer to this question in the following section.

How ML Differs from Software Engineering

Software engineering and ML both solve business problems. Both interact with databases, analyze and code modules, and generate outputs which are used by the business. The business domain understanding is imperative for both fields and so is the usability. On these parameters, both software engineering and ML are similar. However, the key difference lies in the execution and the approach used to solve the business challenge.

Software writing involves writing precise code which can be executed by the processor, that is, the computer. On the other hand, ML collects historical data and understands trends in the data. Based on the trends, the ML algorithm will predict the desired output. Let us look at it with an easy example first.

Consider this: you want to automate the opening of a cola can. Using software, you would code the exact steps with precise coordinates and instructions. For that, you should know those precise details. However, using ML, you would "show" the process of opening a can to the system many times. The system will learn the process by looking at various steps or "train" itself. Next time, the system can open the can itself. Now let's look at a real-life example.

Imagine you are working for a bank which offers credit cards. You are in the fraud detection unit and it is your job to classify a transaction as fraudulent or genuine. Of course, there are acceptance criteria like transaction amount, time of transaction, mode of transaction, city of transaction, and so on.

Let us implement a hypothetical solution using software; you might implement conditions like those depicted in Figure 1-4. Like a decision tree, a final decision can be made. Step 1: if the transaction amount is below the threshold X, then move to step 2 or else accept it. In step 2, the transaction time might be checked and the process will continue from there.

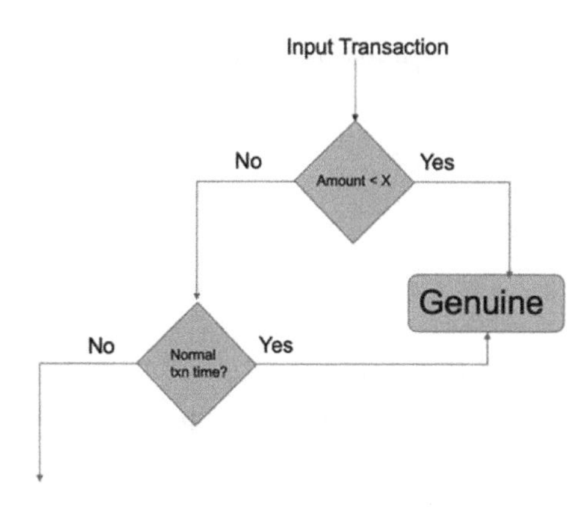

***Figure 1-4.** Hyphothetical software engineering process for a fraud detection system. Software engineering is different from ML.*

However using ML, you will collect the historical data comprising past transactions. It will contain both fraudulent and genuine transactions. You will then expose these transactions to the statistical algorithm and train it. The statistical algorithm will uncover the relationship between attributes of the transaction with its genuine/fraud nature and will keep that knowledge safe for further usage.

Next time, when a new transaction is shown to the system, it will classify it fraudulent or genuine based on the historical knowledge it has generated from the past transactions and the attributes of this new unseen transaction. Hence, the set of rules generated by ML algorithms are dependent on the trends and patterns and offer a higher level of flexibility.

Development of an ML solution is often more iterative than software engineering. Moreover, it is not exactly accurate like software is. But ML is a good generalized solution for sure. It is a fantastic solution for complex business problems and often the only solution for really complicated problems which we humans are unable to comprehend. Here ML plays a pivotal role. Its beauty lies in the fact that if the training data changes, one need not start the development process from scratch. The model can be retrained and you are good to go!

So ML is undoubtedly quite useful, right! It is time for you to understand the steps in an ML project. This will prepare you for a deeper journey into ML.

ML Projects

An ML project is like any other project. It has a business objective to be achieved, some input information, tools and teams, desired accuracy levels, and a deadline!

However, execution of an ML project is quite different. The very first step in the ML process is the same, which is defining a business objective and a measurable parameter for measuring the success criteria. Figure 1-5 shows subsequent steps in an ML project.

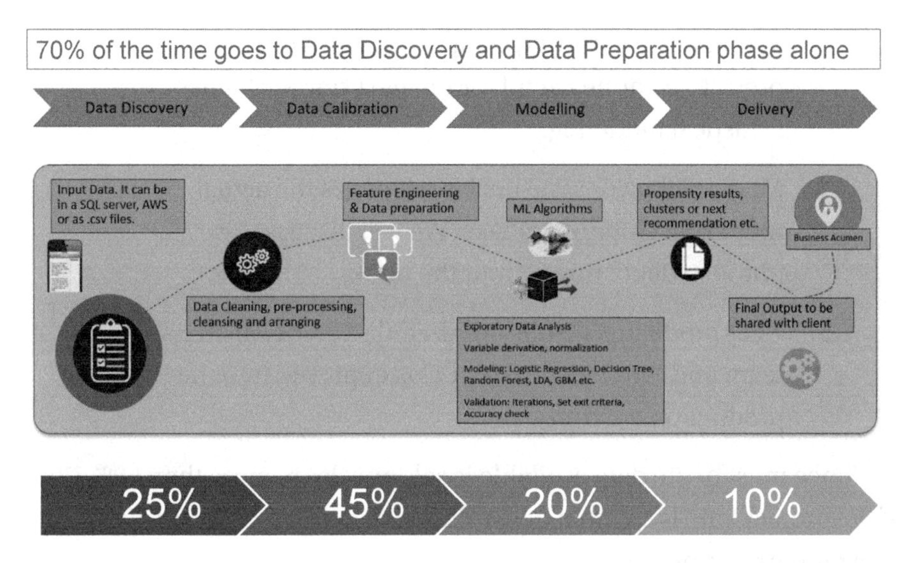

***Figure 1-5.** An ML project is like any other project, with various steps and process. Proper planning and execution are required for an ML project like any other project.*

The subsequent steps are

1. Data discovery is done to explore the various data sources which are available to us. Dataset might be available in SQL server, excel files, text or .csv files, or on a cloud server.

2. In the data mining and calibration stage, we extract the relevant fields from all the sources. Data is properly cleaned and processed and is made ready for the next phase. New derived variables are created and variables which do not have much information are discarded.

3. Then comes the exploratory data analysis or EDA stage. Using analytical tools, general insights are generated from the data. Trends, patterns, and

anomalies are the output of this stage, which prove to be quite useful for the next stage, which is statistical modeling.

4. ML modeling or statistical modeling is the actual model development phase. We will discuss this phase in detail throughout the book.

5. After modeling, results are shared with the business team and the statistical model is deployed into the production environment.

Since most of the data available is seldom clean, more than 60%–70% of the project time is spent in data mining, data discovery, cleaning, and data preparation phase.

Before starting the project, there are some anticipated challenges. In Figure 1-6, we discuss a few questions we should ask before starting an ML project.

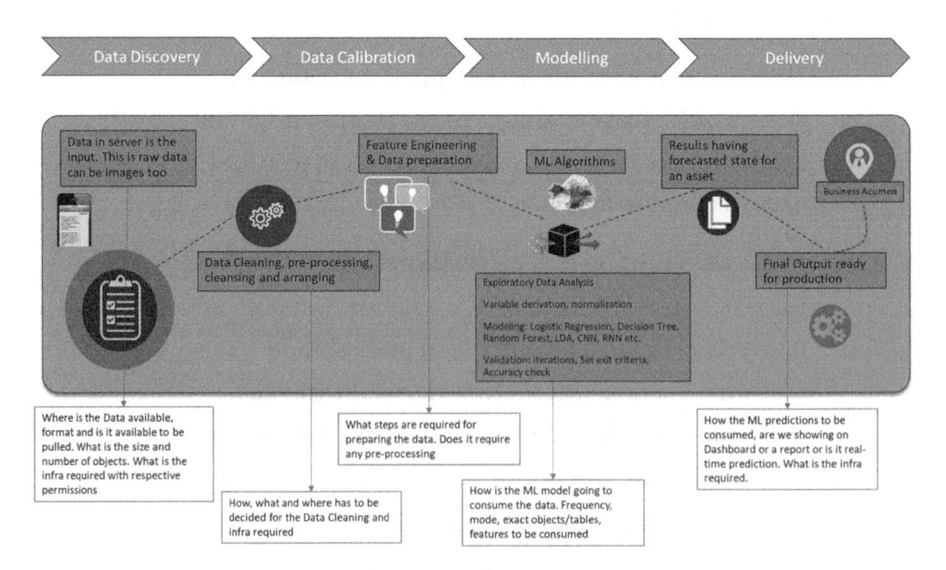

Figure 1-6. *Preparations to be made before starting an ML project. It is imperative that all the relevant questions are clear and KPIs are frozen.*

We should be able to answer these questions about the data availability, data quality, data preparation, ML model prediction measurements, and so on. It is imperative to find the answers to these questions before kicking off the project; else we are risking stress for ourselves and missing deadlines at a later stage.

Now you know what is ML and the various phases in an ML project. It will be useful for you to envisage an ML model and what the various steps are in the process. Before going deeper, it is imperative that we brush up on some statistical and mathematical concepts. You will also agree that statistical and mathematical knowledge is required for you to appreciate ML.

Statistical and Mathematical Concepts for ML

Statistics and mathematics are of paramount importance for complete and concrete knowledge of ML. The mathematical and statistical algorithms used in making the predictions are based on concepts like linear algebra, matrix multiplications, concepts of geometry, vector-space diagrams, and so on. Some of these concepts you would have already studied. While studying the algorithms in subsequent chapters, we will be studying the mathematics behind the working of the algorithms in detail too.

Here are a few concepts which are quite useful and important for you to understand. These are the building blocks of data science and ML:

- **Population vs. Sample**: As the name suggests, when we consider all the data points available to us, we are considering the entire population. If a percentage is taken from the population, it is termed as a sample. This is seen in Figure 1-7.

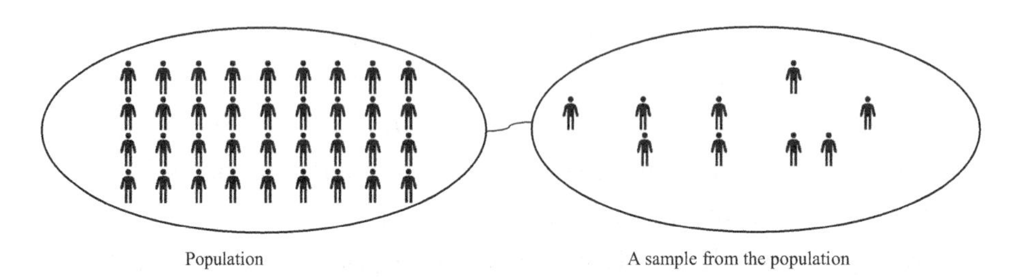

***Figure 1-7.** Population vs. a sample from the population. A sample is a true representation of a population. Sampling should be done keeping in mind that there is no bias.*

- **Parameter vs. Statistic**: Parameter is a descriptive measure of the population: for example, population mean, population variance, and so on. A descriptive measure of a sample is called a statistic. For example, sample mean, sample variance, and so on.
- **Descriptive vs. Inferential Statistics**: When we gather the data about a group and reach conclusions about the same group, it is termed descriptive statistics. However, if data is gathered from a sample and statistics generated are used to generate conclusions about the population from which the sample has been taken, it is called inferential statistics.
- **Numeric vs. Categorical Data**: All data points which are quantitative are numeric, like height, weight, volume, revenue, percentages returns, and so on.
    - The data points which are qualitative are categorical data points: for example, gender, movie ratings, pin codes, place of birth, and so on. Categorical variables are of two types: *nominal* and *ordinal.* Nominal variables do not have a rank between distinct values, whereas ordinal variables have a rank.

- Examples of nominal data are gender, religion, pin codes, ID number, and so on. Examples of ordinal variables are movie ratings, Fortune 50 ranking, and so on.

- **Discrete vs. Continuous Variable**: Data points which are countable are discrete; otherwise data is continuous (Figure 1-8).

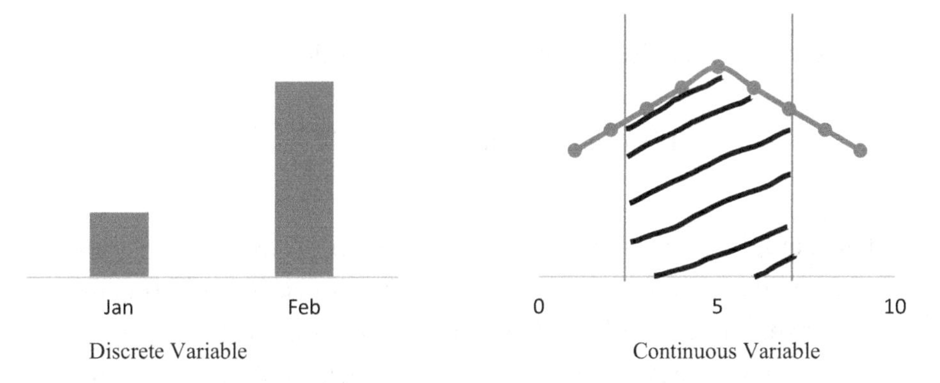

***Figure 1-8.** Discrete variables are countable while continuous variables are in a time frame*

For example, the number of defects in a batch is countable and hence discrete, whereas time between customer arrivals at a retail outlet is continuous.

- **Measures of Central Tendency**: Mean, median, mode, standard deviation, and variance are the measures of central tendency. These measures are central to measuring the various KPIs. There are other measures too which are tracked like total, decile, or quartile distribution. For example, while reporting the number of transactions done in a day, we will report the total number of transactions in a day and the average number per day. We will also report time/date movement for the KPI.

- **Poisson's Distribution**: Poisson's distribution determines the probability of occurrences of a given number of events in a fixed interval of time or space. The assumption is that these events are independent of each other and occur with constant mean.

The equation of Poisson's distribution is as follows:

$$P(\text{k events in the interval}) = \frac{\lambda^k e^{-\lambda}}{k!}$$

For example: if we want to model the number of customers visiting a store between 4 PM and 5 PM or the number of transactions hitting the server between 11 PM and 4 AM, we can use Poisson's distribution.

You can generate Poisson's distribution using the following Python code:

```
import numpy as np
import matplotlib.pyplot as plt
s = np.random.poisson(5, 10000)
count, bins, ignored = plt.hist(s, 14, normed=True)
plt.show()
```

- **Binomial Distribution**: We use binomial distribution to model the number of successes in a sample "n" drawn from population "N." The condition is this sample should be drawn with replacement. Hence, in a sequence of "n" independent events, a Boolean result is decided for success of each of the events. Obviously if the probability of success is "p," then the probability of failure is "1-p".

The equation of binomial distribution is as follows:

$$P(X) = {}^{n}C_{x}\, p^{x} (1-p)^{n-x}$$

The easiest example for a binomial distribution is a coin toss. Each of the coin toss events is independent from the others.

You can generate binomial distribution using the following Python code:

```
import numpy as np
import matplotlib.pyplot as plt
n, p = 10, .5
s = np.random.binomial(n, p, 1000)
count, bins, ignored = plt.hist(s, 14, normed=True)
plt.show()
```

- **Normal or Gaussian Distribution**: Normal distribution or the Gaussian distribution is the most celebrated distribution. It is the famous bell curve and occurs often in nature.

For normal distribution, it is said that when a large number of small, random disturbances influence together then the distribution takes this format. And each of these small disturbances have their own unique distributions.

Gaussian distribution forms the basis of the famous 68–95–99.7 rule, which states that in normal distribution, 68.27%, 95.45%, and 99.73% of the values lie within the one, two, and three standard deviations of the mean. The following Python code is used to generate a normal distribution as shown in Figure 1-9.

```
import numpy as np
import matplotlib.pyplot as plt
mu, sigma = 0, 0.1
s = np.random.normal(mu, sigma, 1000)
```

```
count, bins, ignored = plt.hist(s, 30, normed=True)
plt.plot(bins, 1/(sigma * np.sqrt(2 * np.pi)) * np.exp
( - (bins - mu)**2 / (2 * sigma**2) ),linewidth=2, color='r')
plt.show()
```

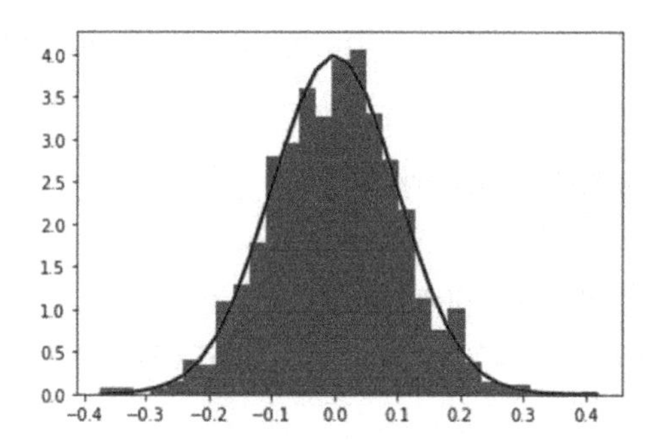

Figure 1-9. *The normal distribution curve is the most celebrated curve*

- **Bias-Variance Trade-off**: To measure how a model is performing we calculate the error, which is the difference between actual and predicted values. This error can occur from two sources, bias and variance, which are shown in Figure 1-10 and defined in Table 1-2.

Error can be represented as follows:

$$\text{Error} = \text{Bias}^2 + \text{Variance} + \text{Irreducible error}$$

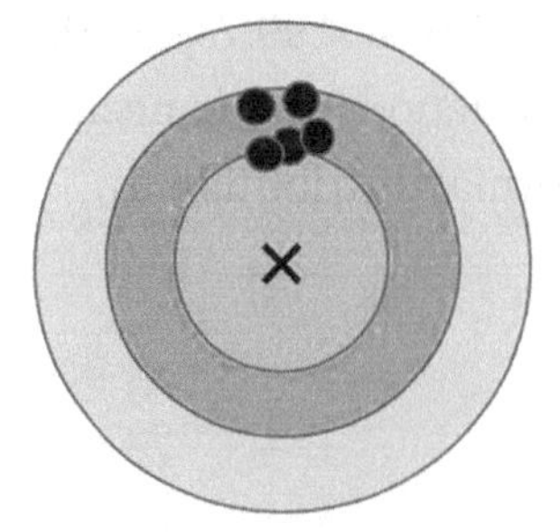

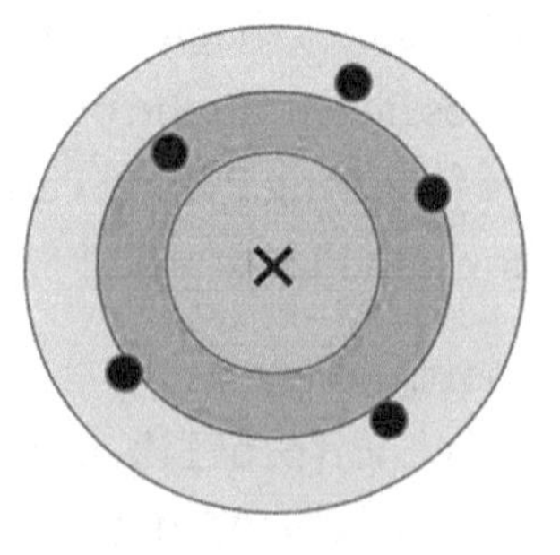

Figure 1-10. *Bias is underfitting of the model, while variance is overfitting of the model. Both have to be managed for a robust ML solution.*

Table 1-2. *Comparison Between Bias and Variance*

Bias	Variance
The measure of how far off the predictions are from the actual value.	The measure of how different the predictions are for only data point.
Bias should be low for a good model.	Variance should be low for a good model.
High bias occurs due to wrong assumptions made during training or underfitting the model by not accounting for all the information present in the data.	High variance occurs when the model is overfitted with the training data and a general rule is not derived. Thus it performs badly with new datasets for predictions.

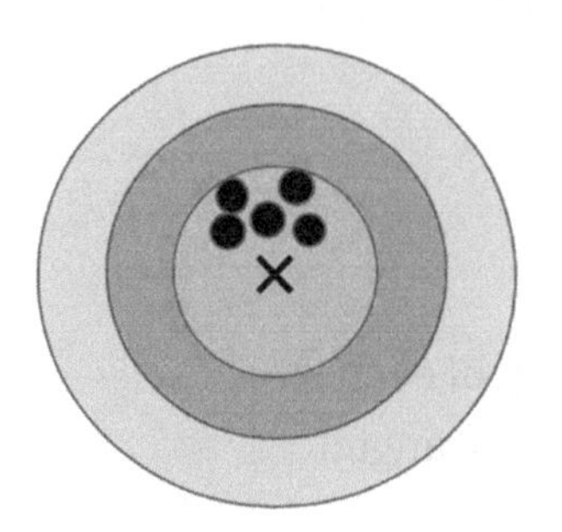

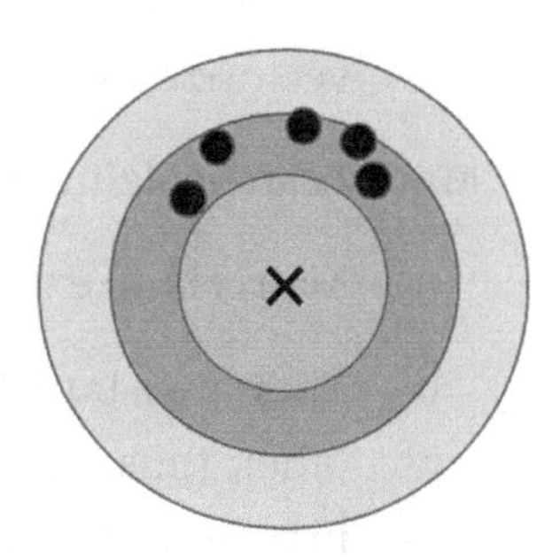

Figure 1-11. *Low variance/low bias and high variance/high bias. Low variance and low bias are desirable from the final shortlisted model.*

- **Vector and Matrix**: The datasets we use can be represented in a vector-space diagram. Hence, let's briefly visit the definition of vector and matrix.

Vector can be defined as follows:

a) Vector is an object which has both magnitude and direction.

b) They are an excellent tool for representing any form of data in mathematical form.

c) A typical vector will look like this: [1,2,3,4,5].

d) A vector in mathematical terms is represented as $\vec{v}$ with an arrow at the top.

e) It can be used for both numerical and non-numerical data. Representation of unstructured data in a mathematical format is achieved through vectorizations and embeddings.

Matrix can be defined as follows:

a) Matrix is an extension of vectors.

b) They are a bunch of vectors placed on top of each other; thus, a matrix consists of numbers arranged in rows and columns.

c) They are a very easy way of representing and holding datasets to perform mathematical operations.

d) A typical matrix will look like this:

$$A = \begin{pmatrix} 1 & 2 & 3 \\ 4 & 5 & 6 \\ 7 & 8 & 9 \end{pmatrix}$$

- **Correlation and Covariance**: Correlation and covariance are very important measures when we try to understand the relationships between variables.

    a) Covariance and correlation are a measure of dependence between two variables.

    b) For example, for a child, as height increases weight generally increases. In this case height and weight are positively correlated.

    c) There can be negative and zero correlation between data points.

    d) For example an increase in absences to class may decrease grades. If the same trend can be observed over a collection of samples, then these parameters are negatively correlated.

    e) Zero correlation shows no linear dependence but there can be nonlinear dependencies. For example, an increase in the price of rice has zero correlation with an increase/decrease in the price of mobiles.

    f) Correlation is the scaled value of covariance.

    g) The three types of correlation—positive, negative, and no-correlation—are shown in Figure 1-12.

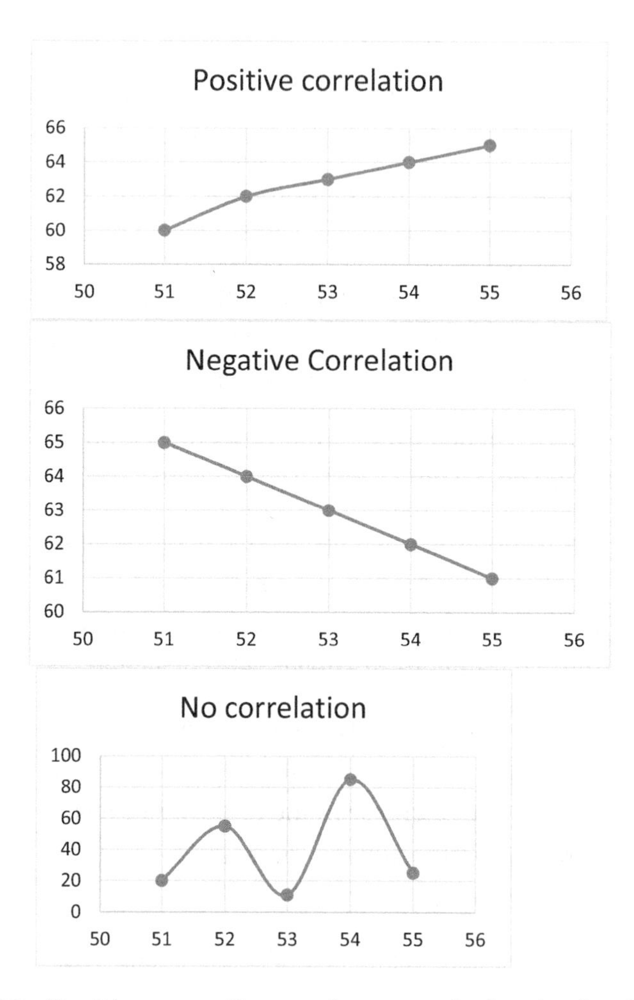

***Figure 1-12.** Positive, negative, and no correlation in the data*

There are still few concepts like measuring the accuracy of the algorithms, R^2, adjusted R^2, AIC values, concordance ratio, KS value, and so on, which are to be discussed. We will be discussing them in two parts: for regression problems in Chapter 2 and for classification problems in Chapter 3.

Great job! Now you have brushed up the major statistical concepts. ML and statistics go hand in hand so kudos on it.

It is now time for you to dive deep into ML concepts, and we will start with different types of algorithms. There are different types of ML algorithms: supervised learning, unsupervised learning, semi-supervised learning, self-learning, feature learning, and so on. We will examine supervised learning algorithms first.

Supervised Learning Algorithms

Supervised learning is arguably the most common usage of ML. As you know, in ML, statistical algorithms are shown historical data to learn the patterns. This process is called *training* the algorithm. The historical data or the *training* data contains both the input and output variables. It contains a set of training examples which are learned by the algorithm.

During the training phase, an algorithm will generate a relationship between output variable and input variables. The goal is to generate a mathematical equation which is capable of predicting the output variable by using input variables. Output variables are also called *target* variables or *dependent* variables, while input variables are called *independent* variables.

To understand this process, let us take an example. Suppose we want to predict the expected price of a house in pounds based on its attributes. House attributes will be its size in sq. m, location, number of bedrooms, number of balconies, distance from the nearest airport, and so on. And for this model we have some historical data at our disposal as shown in Table 1-3. This data is the training data used to train the model.

Table 1-3. *Structure of Dataset to Predict the Price of a House*

Area (sq. m)	Number of bedrooms	Number of balconies	Dist. from airport (km)	Price (mn £)
100	2	0	20	1.1
200	3	1	60	0.8
300	4	1	25	2.9
400	4	2	5	4.5
500	5	2	60	2.5

If the same data is represented in a vector-space diagram it will look like Figure 1-13. Each row or training examples is an *array* or a *feature* vector.

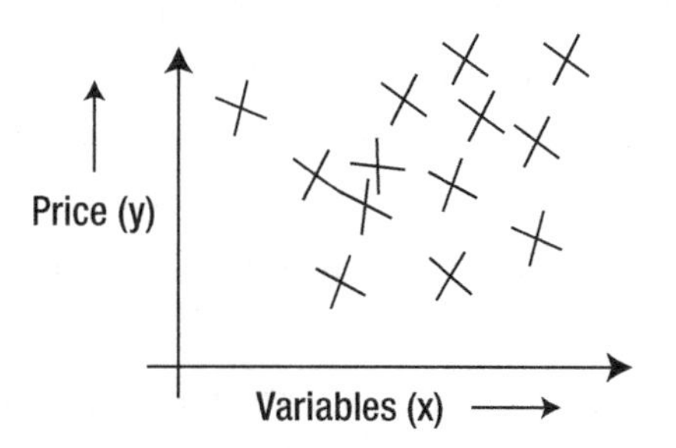

Figure 1-13. *Representation of price and other variables in a vector-space diagram. If we have more than one variable, it can be thought as a multidimensional vector space.*

Now we start the training process. Iteratively, we will try to reach a mathematical function and try to optimize it. The goal is always to improve its accuracy in predicting the house prices.

Basically, what we want to achieve is a function "f" for the price:

price = f (size, location, bedrooms, proximity to city center, balconies ....)

The goal of our ML model is to achieve this equation. Here, price is the target variable and rest are the independent variables. In Figure 1-14, price is our target variable or y, the rest of the attributes are independent variables or x, and the red line depicts the ML equation or the mathematical function. It is also called the *line of best fit*.

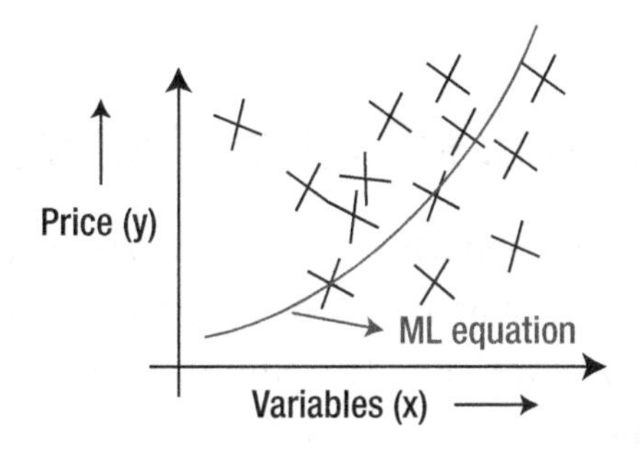

Figure 1-14. *ML equation using regression in vector-space diagram. This equation is the line of best fit used to make the predictions for the unseen dataset.*

The sole aim of this problem is to arrive at this mathematical equation. With more training data points, better and advanced algorithms, and more rigor we constantly strive to improve the accuracy of this equation. This equation is said to be the best representation of our data points. Or using this equation, we can capture the maximum randomness present in the data.

The preceding example is solved using a supervised learning algorithm called linear regression. A different supervised algorithm like a decision tree will require a different approach for the same problem.

Hence, the definition of supervised learning algorithm will be as follows: "Supervised learning algorithms create a statistical ML model to predict the value of a target variable. The input data contains both the independent and the target variable."

The aim of the supervised learning algorithm is to reach an optimized function which is capable of predicting the output associated with new, unseen data. This new, unseen data is not a part of the training data. The function we want to optimize is called the *objective* function.

Supervised learning algorithms are used to solve two kinds of problems: *regression* and *classification*. Let's discuss them now.

Regression vs. Classification Problems

Simply put, regression is used when we want to predict the value of a continuous variable while classification is used when we want to predict the value of a categorical variable. The output of a regression problem will be a continuous value.

Hence, the house prediction example quoted previously is an example of a regression problem, since we want to predict the exact house prices. Other examples are predicting the revenue of a business in the next 30 days, how many customers will make a purchase next quarter, the number of flights landing tomorrow, and how many customers will renew their insurance policies.

On the other hand, let's suppose we want to predict if a customer will churn or not, whether a credit card transaction is fraudulent or not, whether the price will increase or not: these all are examples of *binary* classification problems. If we want to classify between more than two classes, it will be a *multiclass* classification problem. For example, if we want to predict what is going to be the next state of a machine—running, stopped, or paused—we will use a multiclass classification problem. The output of a classification algorithm may be a probability score. So, if we want to decide whether an incoming transaction is fraudulent or not, the classification algorithm can generate a probability score between 0 and 1 for the transaction to be called fraudulent or genuine.

There are quite a few supervised learning algorithms:

1. Linear regression for regression problems
2. Logistic regression for classification problems
3. Decision tree for both regression and classification problems
4. Random forest for both regression and classification problems
5. Support vector machine (SVM) for both regression and classification problems

There are plenty of other algorithms too, like k-nearest neighbor, naive Bayes, LDA, and so on. Neural networks can also be used for both classification and regression tasks. We will be studying all of these algorithms in detail throughout the book. We will be developing Python solutions too.

Tip In general business practice, we compare accuracies of four or five algorithms and select the best ML model to implement in production.

We have examined the definition of supervised learning algorithm and some examples. We will now examine the steps in a supervised learning problem. You are advised to make yourself comfortable with these steps, as will be following them again and again during the entire book.

Steps in a Supervised Learning Algorithm

We discussed steps in an ML project earlier. Here, we will examine the steps specifically for the supervised learning algorithms. The principles of data quality, completeness, and robustness are applicable to each of the steps of a supervised problem.

To solve a supervised learning problem, we follow the steps shown in Figure 1-15. To be noted is that it is an iterative process. Many times, we uncover a few insights that prompt us to go back a step. Or we might realize that an attribute which was earlier thought useful is no longer valid. These iterations are part and parcel of ML, and supervised learning is no different.

Step 1: When we have to solve a supervised learning algorithm, we have the target variable and the independent variables with us. The definition of the target variable is central to solving the supervised learning problem. The wrong definition for it can reverse the results.

For example: if we have to create a solution to detect whether an incoming email is spam or not, the target variable in the training data can be "spam category." If spam category is "1," the email is spam; if it is "0," the email is not spam. In such a case, the model's output will be a probability score for an incoming email to be spam or not. The higher the probability score, higher the chances of the email being spam.

In this step, once we have decided the target variable, we will also determine if it is a regression or a classification problem. If it is a classification problem, we will go one level deeper to identify if it is a binary-classification or multiclass classification problem.

At the end of step 1, we have a target variable defined and a decision on whether it is regression or a classification problem.

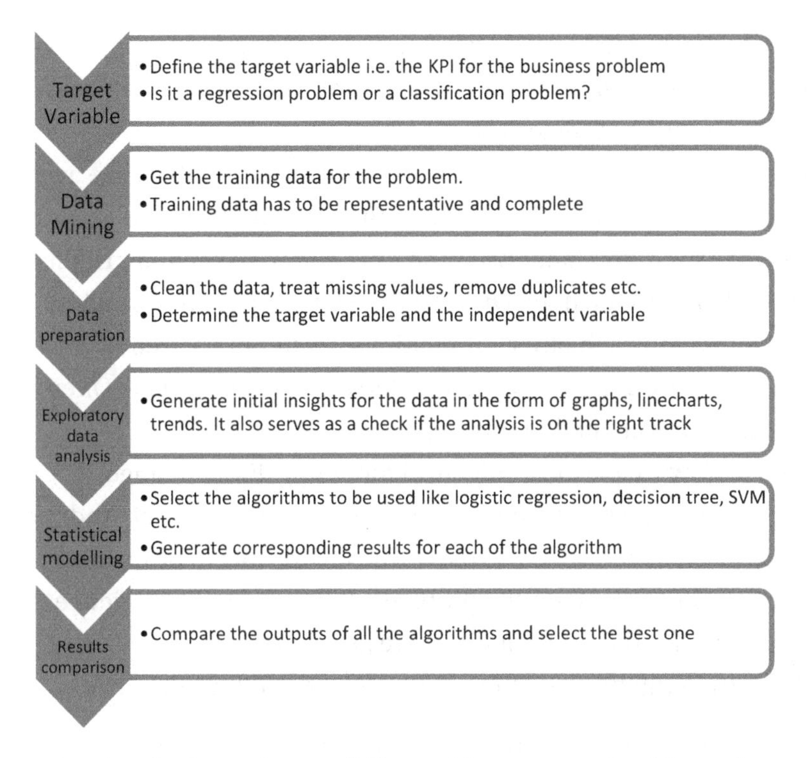

Figure 1-15. *The broad steps followed in a supervised learning algorithm from variable definition to model selection*

Step 2: In the second step, we identify the training data for our model. In this step, best principles regarding the data quality are to be adhered to. The training data consists of both the independent variables and the target variable. This training data will be sourced from all the potential data sources. It should be also representative enough to capture variations from all the time periods to ensure completeness.

Step 3: In this step, data is prepared for statistical modeling. Often, the data is unclean and contains a lot of anomalies. We find null values, NA, NaN, duplicates, and so on in the data. In the date field we might find string values, names might contain integers, and so on, and we have to clean up all the data. In this phase, the target variable is identified, and the independent variables are also known. The independent variables can be categorical or continuous variables. Similarly, the target variable will be either continuous or categorical.

This step also involves creation of new derived variables like average revenue, maximum duration, distinct number of months, and so on.

Step 4: Exploratory analysis is the next step where initial insights are generated from the data. Distribution of independent variables, relationships with each other, co-relations, scatter plots, trends, and so on are generated. We generate a good understanding of the data. Often, many new variables are created in this step too.

Step 5: Now we perform statistical modeling. From a range of supervised learning algorithms, we start by creating a model. And then subsequent algorithms are also used. The algorithms to be used are generally decided based on the problem statement and experience. The accuracy plots for different methods are generated.

While training the algorithm, these steps are followed:

a) The entire data can be split into a 60:20:20 ratio for test, train, and validation data sets. Sometimes, we split into an 80:20 ratio as train and test data.

b) But if the numbers of raw data (for example images) are quite high (1 million), a few studies suggest having 98% train, 1% test, and 1% validation datasets.

c) All three datasets though, should always be randomly sampled from the original raw data with no bias in selection. It is imperative since if the testing or validation datasets are not a true representative of the training data, we will not be measuring the efficacy correctly.

d) However, there can be instances wherein a sampling bias cannot be avoided. For example, if a demand forecasting solution is being modeled, we will use data from the historical time period to train the algorithm. The time dimension will be used while creating training and testing datasets.

e) The training dataset will be used for training the algorithm. The independent variables act as a guiding factor and the target variable is the one which we try to predict.

f) The testing dataset will be used to compare the testing accuracy. Testing/validation data is not exposed to the algorithm during the training phase.

g) We should note that testing accuracy is much more important than training accuracy. Since the algorithm should be able to generalize better on unseen data, the emphasis is on testing accuracy.

h) There are instances where accuracy may not be the KPI we want to maximize. For example, while we are creating a solution to predict if a credit card transaction is fraudulent or not, accuracy is not the target KPI.

i) During the process of model development, we iterate through various input parameters to our model, which are called *hyperparameters*. Hyperparameter tuning is done to achieve the best and most stable solution.

j) The validation dataset should be exposed to the algorithm only once, after we have finalized the network/algorithm and are done with tuning.

Step 6: In this step, we compare and contrast the accuracies we have generated from the various algorithms in step 4. A final solution is the output of this step. It is followed by discussion with the SME and then implementation in the production environment.

These are the broad steps in supervised learning algorithms. These solutions will be developed in great detail in Chapters 2, 3, and 4.

Note Preprocessing steps like normalization are done for training data only and not for validation datasets, to avoid data leakage.

This brings us to the end of discussion on supervised learning algorithms. It is time to focus on other types of ML algorithms, and next in the queue is unsupervised learning.

Unsupervised Learning Algorithms

We know supervised learning algorithms have a target variable which we want to predict. On the other hand, unsupervised algorithms do not have any prelabeled data and hence they look for undetected patterns in the data. This is the key difference between supervised and unsupervised algorithms.

For example, the marketing team of a retailer may want to improve customer stickiness and customer lifetime value, increase the average revenue of the customer, and improve targeting through marketing campaigns. Hence, if the customers can be clubbed and segmented in similar clusters, the approach can be very effective. This problem can be solved using an unsupervised learning algorithm. Unsupervised analysis can be majorly categorized in *cluster analysis* and dimensionality reduction techniques like *principal components analysis (PCA)*. Let's discuss cluster analysis first.

Cluster Analysis

The most famous unsupervised learning application is *cluster analysis.* Cluster analysis groups the data based on similar patterns and common attributes visible in the data. The common patterns can be the presence or absence of those similar features. The point to be noted is that we do not have any benchmark or labeled data to guide; hence the algorithm is finding the patterns. The example discussed previously is a customer segmentation use case using cluster analysis. The customers of the retailer will have attributes like revenue generated, number of invoices, distinct products bought, online/offline ratio, number of stores visited, last transaction date, and so on. When these customers are visualized in a vector-space diagram, they will look like Figure 1-16(i). After the customers have been clustered based on their similarities, the data will look like Figure 1-16(ii).

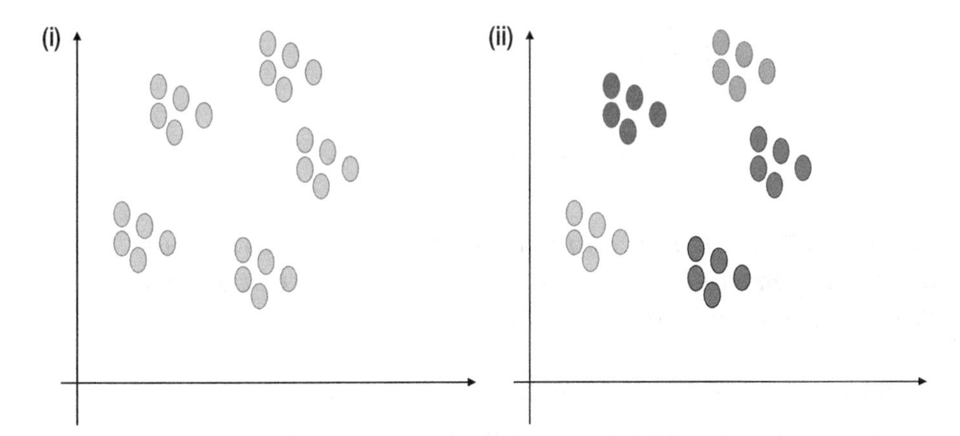

Figure 1-16. *(i) Before clustering of data; (ii) after clustering of data*

There are quite a few clustering algorithms available—*k-means* clustering, *hierarchical* clustering, *DBScan, spectral* clustering, and so on. The most famous and widely used clustering algorithm is k-means clustering.

PCA

In ML and data science, we always strive to make some sense from randomness, and gather insights from haphazard data sources. Recall from the supervised learning algorithm discussion and Figure 1-9 that we have represented the line of best fit, that is, the mathematical equation which is able to capture the maximum randomness present in the data. This randomness is captured using the various attributes or independent variables. But imagine if you have 80 or 100 or 500 such variables. Won't it be a tedious task? Here PCA helps us.

Refer to Figure 1-17. The two principal components, PC 1 and PC 2, are orthogonal to each other and capturing the maximum randomness in the data. That is PCA.

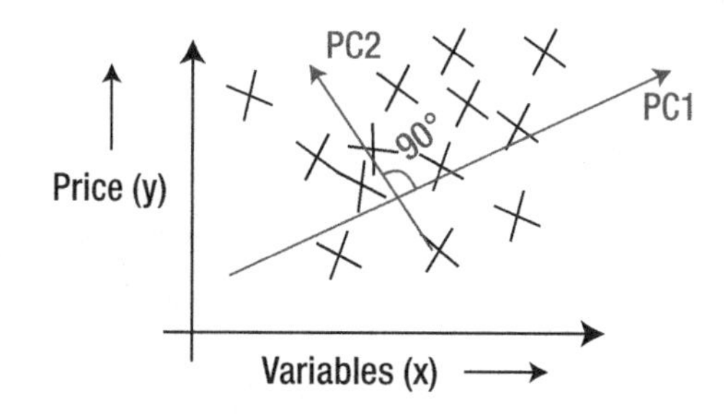

Figure 1-17. *Principal components, PC1 and PC2, to capture the maximum randomness in the data. PCA is quite a popular dimensionality reduction technique.*

So, in PCA we define the randomness in the data by a principal component which is able to capture the maximum variation. The next principal component is orthogonal to the first one so that it can capture maximum variation, and so on. Hence, PCA serves as a dimensionality reduction solution for us. Instead of requiring all the attributes, we use the principal components and use them instead of all the attributes.

Now, we will examine the semi-supervised type of ML algorithms.

Semi-supervised Learning Algorithms

Semi-supervised algorithms can be called a combination of supervised and unsupervised learning algorithms; or, they fall between the two. When we have a small set of labeled data and a large amount of unlabeled data, semi-supervised algorithms help us in resolving the issue.

The assumption in semi-supervised learning is that the data points which belong to the same cluster or group will tend to have the same label. Hence, once unsupervised algorithms like k-means clustering will share the output clusters, using labeled data we can further improve the quality of the data.

Semi-supervised learning algorithms are used in use cases either where labeled data is not generated or where labeling the target variable is going to be a time-consuming and costly affair. Generative models, graph-based methods, and low-density separation are some of the methods used in semi-supervised learning algorithms.

This marks the end of discussing major types of ML algorithms. There are other families of algorithms like association rule–based market basket analysis, r*einforcement learning, and so on*. You are advised to explore them too.

In the next section, we will go through the available list of technical tools which help us in data management, data analysis, ML, and visualizations.

Technical Stack

Tools are an integral part of data science and ML. Tools are required for data management, data mining, analysis, and building the actual ML model.

A brief list of the various utilities, languages, and tools follows:

- **Data Engineering**: Spark, Hadoop, SQL, Redshift, Kafka, Java, C++

- **Data Analysis**: Excel, SQL, postgres, mySQL, NoSQL, R, Python
- **ML**: SAS, R, Python, Weka, SPSS, MATLAB
- **Visualizations**: Tableau, PowerBI, Qlik, COGNOS
- **Cloud Services**: Microsoft Azure, Amazon Web Services, Google Cloud Platform

These tools or rather a combination of them is used for the complete project—starting from data management and data mining to ML and to visualization.

Tip All tools are good to be used and will generate similar results. The choice is often between open source and licensed or how scalable the solution is envisioned to be.

These tools act as a building block for the project. You are advised to get some level of understanding for each of the components. It will be helpful to appreciate all the facets of a data science project.

While making a choice of an ML tool, to arrive at the best solution suite, we should consider the following parameters too:

- **Ease of Deployment**: how easy it is to deploy the model in the production environment
- **Scalability**: whether the solution is scalable to other products and environment
- **Maintenance and Model Refresh**: ease of maintaining and refreshing the model regularly
- **Speed**: speed of making the predictions; sometimes, the requirement is in real-time

- **Cost** (license and man hours required): what are the license cost and efforts required
- **Support Available**: What type of support is available with us from the team; for example, the MATLAB team extends support since it requires license while Python being open source does not have a support system like MATLAB

You are advised to get a decent level of understanding of at least one or two tools from each of the buckets. SQL and Microsoft Excel are ubiquitous and hence are recommended. Python is a leading tool for ML and AI. With the release of deep learning frameworks like TensorFlow and Keras, Python has generated a huge base of users. And in this book too we are going to use Python only.

We are heading towards the end of the first chapter. We will discuss the reasons for ML being highly used in businesses in the next section.

ML's Popularity

The beauty of ML algorithms lies in their capability to solve complex problems which are otherwise quite difficult for us to solve. We humans can only visualize a few dimensions simultaneously. Using ML algorithms, not only can multiple dimensions be visualized and analyzed together, trends and anomalies can be detected easily.

Using ML we can work on complex data like images and text that are quite difficult to analyze otherwise. ML and particularly deep learning allows us to create automated solutions.

Here are some factors which are playing a vital role in making ML popular:

1) **Interest by the business**: Now businesses and stakeholders have renewed interest in harnessing the power of data and implementing ML. Data science departments are set up in the organizations, and there are dedicated teams which are leading the discussions with various processes. We have also witnessed a surge in the number of startups entering the field.

2) **Computation power**: The computation power now available to us is huge as compared to a few decades back. It is also cheaply available. GPU and TPU are making the computations faster and now we have repositories to store terabytes of data. Cloud-based computations are making the process faster. We now have Google Colaboratory to run the codes by using the excellent computations available there.

3) **Data explosion**: The amount of data available to us has increased exponentially. With more social media platforms and interactions and the world going online and virtual, the amount of data generated is across domains. Now, more and more businesses and processes are capturing time-based attributes and creating system dynamics and virtualizations to capture the data points. We now have more structured data points stored in ERP, SAP systems. More videos are uploaded to YouTube, photos are uploaded to FaceBook and Instagram, and text news is flowing across the globe—all refer to zettabytes of data getting generated and ready for analysis.

4) **Technological advancements**: Now we have more sophisticated statistical algorithms at our disposal. Deep learning is pushing the limits further and further. Great emphasis and effort are now put into data engineering, and with emerging technologies we are constantly evolving and improving the efficiency and accuracy of the systems.

5) **Availability of the human capital**: There is an increased interest in mastering data science and ML; hence the number of data analysts and data scientists is increasing.

These are some of the factors making ML one of the most sought emerging technologies. And indeed it is delivering fantastic results too across domains and processes. A few of the important ones are listed in the next section, where we discuss the uses of ML.

Use Cases of ML

ML is getting its application across domains and business. We are sharing a few use cases already implemented in the industry. This is not an exhaustive list but only few:

- **Banking, financial services, and insurance sector**: The BFSI sector is quite advanced in implementing ML and AI. Throughout the value chain, multiple solutions have been implemented, a few of which follow:
    - Credit card fraud detection model is used to classify credit card transactions as fraudulent or genuine. Rule-based solutions do exist, but ML models strengthen the capability further. Supervised classification algorithms can be used here.

  - Cross-sell and up-sell products: this allows banks and insurance companies to increase the product ownership by the customers. Unsupervised clustering can be used to segment the customers followed by targeted marketing campaigns.
  - Improve customer lifetime value and increase the customer retention with the business. Customer propensity models can be built using supervised classification algorithms.
  - Identify potential insurance premium defaulters using supervised classification algorithms.
- **Retail**: Grocery, apparel, shoes, watches, jewelry, electronic retail, and so on are utilizing data science and ML in a number of ways. A few examples follow:
  - Customer segmentation using unsupervised learning algorithms is done to improve the customer engagement. Customer's revenue and number of transactions can be improved by using targeted and customized marketing campaigns.
  - Demand forecasting is done for better planning using supervised regression methods. Pricing models are being developed to price the good in an intelligent manner too.
  - Inventory optimization is done using ML, leading to improved efficiency and decrement in the overall costs.

- Customer churn propensity models predict which customers are going to churn in the next few months. Proactive action can be taken to save these customers from churn. Supervised classification algorithms help in creating the model.
- Supply chain optimization is done using data science and ML to optimize the inventory.

- **Telecommunication**: The telecom sector is no different and is very much ahead in using data science and ML. Some use cases are as follows:
    - Customer segmentation to increase the ARPU (average revenue per user) and VLR (visitor location register) using unsupervised learning algorithms
    - Customer retention is improved using customer churn propensity models using supervised classification algorithms
    - Network optimization is done using data science and ML algorithms
    - Product recommendation models are used to recommend the next best prediction and next best offer to customers based on their usage and behavior
- **Manufacturing industry**: The manufacturing industry generates a lot of data in each and every process. Using data science and ML, a few use cases follow:
    - Predictive maintenance is done using supervised learning algorithms; this avoids breakdown of the machines and proactive action can be taken

- Demand forecasting is done for better planning and resource optimization
- Process optimization is done to identify bottlenecks and reduce overheads
- Identification of tools and combinations of them to generate the best possible results are predicted using supervised learning algorithms

There are multiple other domains and avenues where ML is getting implemented like aviation, energy, utilities, health care, and so on. AI is opening new capabilities by implementing speech-to-text conversion in call centers, object detection and tracking is done in surveillance, image classification is done to identify defective pieces in manufacturing, facial recognition is used for security systems and crowd management, and so on.

ML is a very powerful tool and should be used judiciously. It can help us automate a lot of processes and enhance our capabilities multifold.

With that, we are coming to the end of the first chapter. Let's proceed to the summary now!

Summary

Data is changing our approach towards the decision-making process. More and more business decisions are now data-driven. Be it marketing, supply chain, human resources, pricing, product management—there is no business process left untouched. And data science and ML are making the journey easier.

ML and AI are rapidly changing our world. Trends are closely monitored, anomalies are detected, alarms are raised, aberrations from the normal process are witnessed, and preventive actions are being taken. Preventive actions are resulting in cost saving, optimized processes, saving of time and resources, and in some instances saving life too.

It is imperative to assure that data available to use is of good quality. It defines the success factor of the model. Similarly, tools often play a vital role in final success or failure. Organizations are looked up to for their flexibility and openness to using newer tools.

It is also necessary that due precaution is taken while we are designing the database, conceptualizing the business problem, or finalizing the team to deliver the solution. It requires a methodical process and a deep understanding of the business domain. Business process owners and SMEs should be an integral part of the team.

In this first chapter, we have examined various types of ML, data and attributes of data quality, and ML processes. With this, you have gained significant knowledge of the data science domain. We also learned about various types of ML algorithms, their respective use cases and examples, and how they can be used for solving business problems. This chapter serves as a foundation and stepping-stone for the next chapters. In the next chapter, we are going to start focusing on supervised learning algorithms and we will be starting with regression problems. The various types of regression algorithms, the mathematics, pros and cons, and Python implementation will be discussed.

You should be able to answer these questions.

EXERCISE QUESTIONS

Question 1: What is ML and how is it different from software engineering?

Question 2: What are the various data types available and what are the attributes of good-quality data?

Question 3: What are the types of ML algorithms available to be used?

Question 4: What is the difference between Poisson's, binomial, and normal distributions and what are some examples of each?

Question 5: What are the steps in a supervised learning algorithm?

Question 6: Where is ML being applied?

Question 7: What are the various tools available for data engineering, data analysis, ML, and data visualizations?

Question 8: Why is ML getting so popular and what are its distinct advantages?

CHAPTER 2

Supervised Learning for Regression Analysis

"The only certainty is uncertainty."

— Pliny the Elder

The future has uncertainty for sure. The best we can do is plan for it. And for meticulous planning, we need to have a peek into the future. If we can know beforehand how much the expected demand for our product is, we will manufacture adequately—not more, not less. We will also rectify the bottlenecks in our supply chain if we know what the expected traffic of goods is. Airports can plan resources better if they know the expected inflow of passengers. Or ecommerce portals can plan for the expected load if they know how many clicks are expected during the upcoming sale season.

It may not be possible to forecast accurately, but it is indeed required to predict these values. It is still done, with or without ML-based predictive modeling, for financial budget planning, allocation of resources, guidance issued, expected rate of growth, and so on. Hence, the estimation of such values is of paramount importance. In this second chapter, we will be studying the precise concepts to predict such values.

V. Verdhan, *Supervised Learning with Python*,
https://doi.org/10.1007/978-1-4842-6156-9_2

In the first chapter, we introduced supervised learning. The differentiation between supervised, unsupervised, and semi-supervised was also discussed. We also examined the two types of supervised algorithms: regression and classification. In this second chapter, we will study and develop deeper concepts of supervised regression algorithms.

We will be examining the regression process, how a model is trained, behind-the-scenes process, and finally the execution for all the algorithms. The assumptions for the algorithms, the pros and cons, and statistical background beneath each one will be studied. We will also develop code in Python using the algorithms. The steps in data preparation, data preprocessing, variable creation, train-test split, fitting the ML model, getting significant variables, and measuring the accuracy will all be studied and developed using Python. The codes and datasets are uploaded to a GitHub repository for easy access. You are advised to replicate those codes yourself.

Technical Toolkit Required

We are going to use Python 3.5 or above in this book. You are advised to get Python installed on your machine. We will be using Jupyter notebook; installing Anaconda-Navigator is required for executing the codes. All the datasets and codes have been uploaded to the Github repository at `https://github.com/Apress/supervised-learning-w-python/tree/master/Chapter2` for easy download and execution.

The major libraries used are numpy, pandas, matplotlib, seaborn, scikit learn, and so on. You are advised to install these libraries in your Python environment.

Let us go into the regression analysis and examine the concepts in detail!

Regression analysis and Use Cases

Regression analysis is used to estimate the value of a *continuous* variable. Recall that a continuous variable is a variable which can take any numerical value; for example, sales, amount of rainfall, number of customers, number of transactions, and so on. If we want to estimate the sales for the next month or the number of passengers expected to visit the terminal in the next week or the number of customers expected to make bank transactions, we use regression analysis.

Simply put, if we want to predict the value of a continuous variable using supervised learning algorithms, we use regression methods. Regression analysis is quite central to business solving and decision making. The predicted values help the stakeholders allocate resources accordingly and plan for the expected increase/decrease in the business.

The following use cases will make the usage of regression algorithms clear:

1. A retailer wants to estimate the number of customers it can expect in the upcoming sale season. Based on the estimation, the business can plan on the inventory of goods, number of staff required, resources required, and so on to be able to cater to the demand. Regression algorithms can help in that estimation.

2. A manufacturing plant is doing a yearly budget planning. As a part of the exercise, expenditures under various headings like electricity, water, raw material, human resources, and so on have to be estimated in relation to the expected demand. Regression algorithms can help assess historical data and generate estimates for the business.

3. A real estate agency wishes to increase its customer base. One important attribute is the price of the apartments, which can be improved and generated judiciously. The agency wants to analyze multiple parameters which impact the price of property, and this can be achieved by regression analysis.

4. An international airport wishes to improve the planning and gauge the expected traffic in the next month. This will allow the team to maintain the best service quality. Regression analysis can help in that and make an estimation of the number of estimated passengers.

5. A bank which offers credit cards to its customers has to identify how much credit should be offered to a new customer. Based on customer details like age, occupation, monthly salary, expenditure, historical records, and so on, a credit limit has to be prescribed. Supervised regression algorithms will be helpful in that decision.

There are quite a few statistical algorithms to model for the regression problems. The major ones are listed as

1. Linear regression
2. Decision tree
3. Random forest
4. SVM
5. Bayesian methods
6. Neural networks

We will study the first three algorithms in this chapter and the rest in Chapter 4. We are starting with linear regression in the next section.

What Is Linear Regression

Recall from the section in Chapter 1 where we discussed the house price prediction problem using area, number of bedrooms, balconies, location, and so on. Figure 2-1(i) represents the representation of data in a vector-space diagram, while on the right in Figure 2-1(ii) we have suggested an ML equation termed as the line of best fit to explain the randomness in the data and predict the prices.

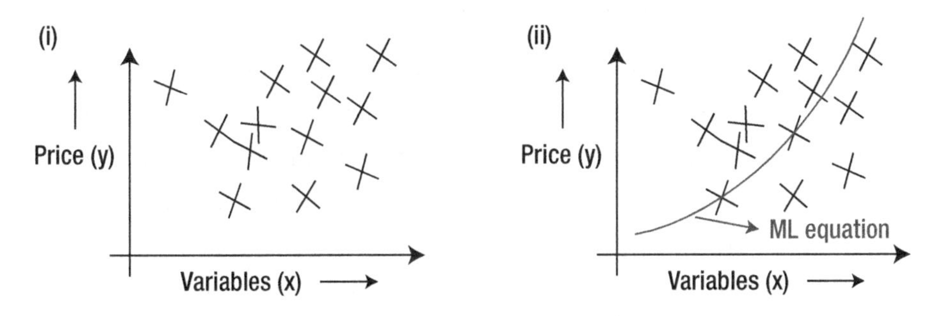

Figure 2-1. *(i) The data in a vector-space diagram depicts how price is dependent on various variables. (ii) An ML regression equation called the line of best fit is used to model the relationship here, which can be used for making future predictions for the unseen dataset.*

In the preceding example, there is an assumption that price is correlated to size, number of bedrooms, and so on. We discussed *correlation* in the last chapter. Let's refresh some points on correlation:

- Correlation analysis is used to measure the strength of association (linear relationship) between two variables.

- If two variables move together in the same direction, they are positively correlated. For example: height and weight will have a positive relationship. If the two variables move in the opposite direction, they are negatively correlated. For example, cost and profit are negatively related.
- The range of correlation coefficient ranges from -1 to +1. If the value is -1, it is absolute negative correlation; if it is +1, correlation is absolute positive.
- If the correlation is 0, it means there is not much of a relationship. For example, the price of shoes and computers will have low correlation.

The objective of the linear regression analysis is to measure this relationship and arrive at a mathematical equation for the relationship. The relationship can be used to predict the values for unseen data points. For example, in the case of the house price problem, predicting the price of a house will be the objective of the analysis.

Formally put, linear regression is used to predict the value of a dependent variable based on the values of at least one independent variable. It also explains the impact of changes in an independent variable on the dependent variable. The dependent variable is also called the target variable or *endogenous* variable. The independent variable is also termed the explanatory variable or *exogenous* variable.

Linear regression has been in existence for a long time. Though there are quite a few advanced algorithms (some of which we are discussing in later chapters), still linear regression is widely used. It serves as a benchmark model and generally is the first step to study supervised learning. You are advised to understand and examine linear regression closely before you graduate to higher-degree algorithms.

Let us say we have a set of observations of x and Y where x is the independent variable and Y is the dependent variable. Equation 2-1 describes the linear relation between x and Y:

$$Y_i = \beta_0 + \beta_1 x_i + \varepsilon_i \qquad \text{(Equation 2-1)}$$

where

Y_i = *Dependent* variable or the target variable which we want to predict

x_i = *Independent* variable or the predicting variables used to predict the value of Y

β_0 = Population Y *intercept*. It represents the value of Y when the value of x_i is zero

β_1 = Population slope *coefficient*. It represents the expected change in the value of Y by a unit change in x_i

ε = *random error* term in the model

Sometimes, β_0 and β_1 are called the *population model coefficients*.

In the preceding equation, we are suggesting that the changes in Y are assumed to be caused by changes in x. And hence, we can use the equation to predict the values of Y. The representation of the data and Equation 2-1 will look like Figure 2-2.

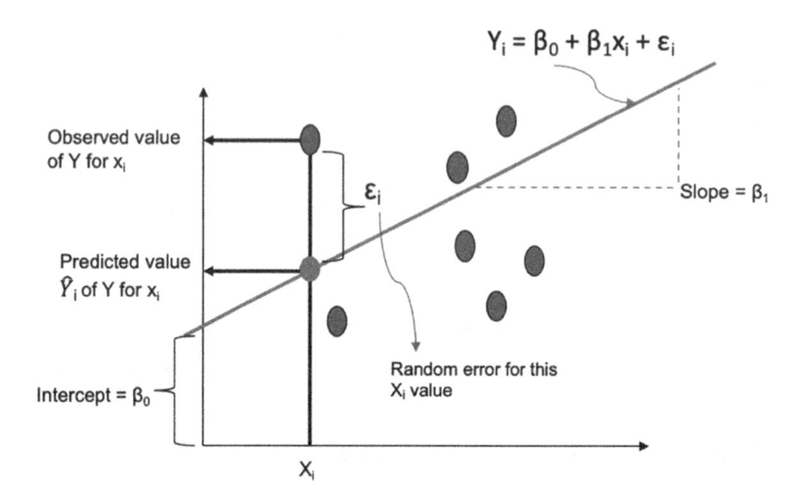

***Figure 2-2.** A linear regression equation depicted on a graph showing the intercept, slope, and actual and predicted values for the target variable; the red line shows the line of best fit*

These model coefficients (β_0 and β_1) have an important role to play. Y intercept (β_0) is the value of dependent variable when the value of independent variable is 0, that is, it is the default value of dependent variable. Slope (β_1) is the slope of the equation. It is the change expected in the value of Y with unit change in the value of x. It measures the impact of x on the value of Y. The higher the absolute value of (β_1), the higher is the final impact.

Figure 2-2 also shows the predicted values. We can understand that for the value of x for observation i, the actual value of dependent variable is Y_i and the predicted value or estimated value is $\hat{Y}_i$.

There is one more term here, *random error*, which is represented by ε. After we have made the estimates, we would like to know how we have done, that is, how far the predicted value is from the actual value. It is represented by random error, which is the difference between the predicted and actual value of Y and is given by $\varepsilon = \left(\hat{Y}_i - Y_i\right)$. It is important to note that the smaller the value of this error, the better is the prediction.

There is one more important point to ponder upon here. There can be multiple lines which can be said to represent the relationship. For example, in the house price prediction problem, we can use many equations to determine the relationship as shown in Figure 2-3(ii) in different colors.

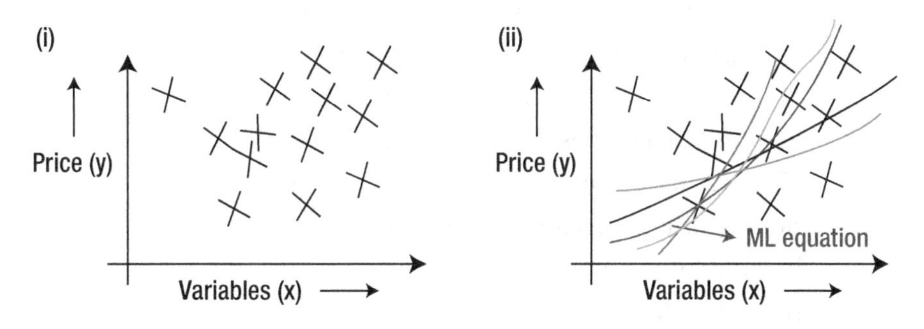

Figure 2-3. *(i) The data in a vector-space diagram depicts how price is dependent on various variables. (ii) While trying to model for the data, there can be multiple linear equations which can fit, but the objective will be to find the equation which gives the minimum loss for the data at hand.*

Hence, it turns out that we have to find out the best mathematical equation which can minimize the random error and hence can be used for making the predictions. This is achieved during training of the regression algorithm. During the process of training the linear regression model, we get the values of β_0 and β_1, which will minimize the error and can be used to generate the predictions for us.

The linear regression model makes some assumptions about the data that is being used. These conditions are a litmus test to check if the data we are analyzing follows the requirements or not. More often, data is *not* conforming to the assumptions and hence we have to take corrective measures to make it better. These assumptions will be discussed now.

Assumptions of Linear Regression

The linear regression has a few assumptions which need to be fulfilled. We check these conditions and depending on the results, decide on next steps:

1. Linear regression needs the relationship between dependent and independent variables to be linear. It is also important to check for outliers since linear regression is sensitive to outliers. The linearity assumption can best be tested with scatter plots. In Figure 2-4, we can see the meaning of linearity vs. nonlinearity. The figure on the left has a somewhat linear relationship where the one on the right does not have a linear relationship.

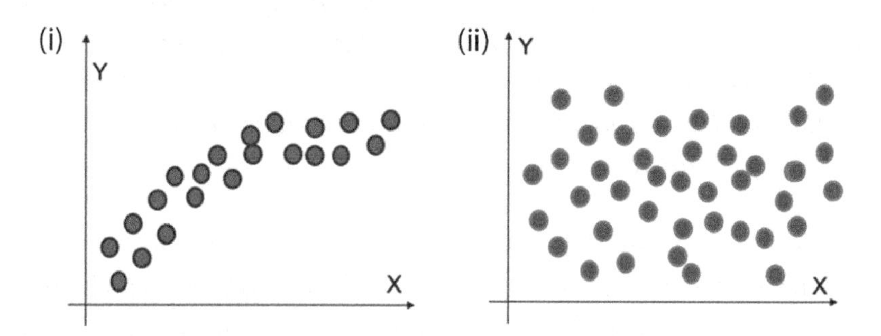

Figure 2-4. *(i) There is a linear relationship between X and Y. (ii) There is not much of a relationship between X and Y variables; in such a case it will be difficult to model the relationship.*

2. Linear regression requires all the independent variables to be multivariate normal. This assumption can be tested using a histogram or a Q-Q plot. The normality can be checked with a goodness-of-fit test like the Kolmogorov-Smirnov

test. If the data is not normally distributed, a nonlinear transformation like log transformation can help fix the issue.

3. The third assumption is that there is little or no *multicollinearity* in the data. When the independent variables are highly correlated with each other, it gives rise to multicollinearity. Multicollinearity can be tested using three methods:

    a. **Correlation matrix**: In this method, we measure Pearson's bivariate correlation coefficient between all the independent variables. The closer the value to 1 or -1, the higher is the correlation.

    b. **Tolerance**: Tolerance is derived during the regression analysis only and measures how one independent variable is impacting other independent variables. It is represented as $(1 - R^2)$. We are discussing R^2 in the next section If the value of tolerance is < 0.1, there are chances of multicollinearity present in the data. If it is less than 0.01, then we are sure that multicollinearity is indeed present.

    c. **Variance Inflation Factor (VIF)**: VIF can be defined as the inverse of tolerance (1/T). If the value of VIF is greater than 10, there are chances of multicollinearity present in the data. If it is greater than 100, then we are sure that multicollinearity is indeed present.

Note Centering the data (deducting mean from each score) helps to solve the problem of multicollinearity. We will examine this in detail in Chapter 5!

---

4. Linear regression assumes that there is little or no *autocorrelation* present in the data. Autocorrelation means that the residuals are not independent of each other. The best example for autocorrelated data is in a time series data like stock prices. The prices of stock on T_{n+1} are dependent on T_n. While we use a scatter plot to check for the autocorrelation, we can use Durbin-Watson's "d-test" to check for the autocorrelation. Its null hypothesis is that the residuals are not autocorrelated. If the value of d is around 2, it means there is no autocorrelation. Generally d can take any value between 0 and 4, but as a rule of thumb $1.5 < d < 2.5$ means there is no autocorrelation present in the data. But there is a catch in the Durbin-Watson test. Since it analyzes only linear autocorrelation and between direct neighbors only, many times, scatter plot serves the purpose for us.

5. The last assumption in a linear regression problem is *homoscedasticity*, which means that the residuals are equal over the regression line. While a scatter plot can be used to check for homoscedasticity, the Goldfeld-Quandt test can also be used for heteroscedasticity. This test divides the data into two groups. Then it tests if the variance of residuals is similar across the two groups. Figure 2-5 shows an example of residuals which are not homoscedastic. In such a case, we can do a nonlinear correction to fix the problem. As we can see in Figure 2-5, heteroscedasticity results in a distinctive cone shape in the residual plots.

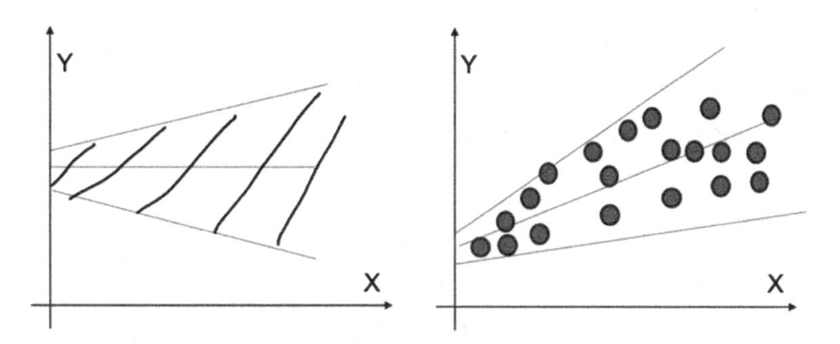

Figure 2-5. *Heteroscedasticity is present in the dataset and hence there is a cone-like shape in the residual scatter plot*

These assumptions are vital to be tested and more often we transform our data and clean it. It is imperative since we want our solution to work very well and have a good prediction accuracy. An accurate ML model will have a low loss. This accuracy measurement process for an ML model is based on the target variable. It is different for classification and regression problems. In this chapter, we are discussing regression base models while in the next chapter we will examine the classification model's accuracy measurement parameters. That brings us to the important topic of measuring the accuracy of a regression problem, which is the next topic of discussion.

Measuring the Efficacy of Regression Problem

There are different ways to measure the robustness of a regression problem. The *Ordinary least-squares (OLS) method* is one of the most used and quoted ones. In this method, β_0 and β_1 are obtained by finding their respective values which minimize the sum of the squared distance between Y and $\hat{Y}$, which are nothing but the actual and predicted values of the dependent variable as shown in Figure 2-6. This is referred to as the *loss function*, which means the loss incurred in making a prediction. Needless to say, we want to minimize this loss to get the best model. The errors are referred as *residuals* too.

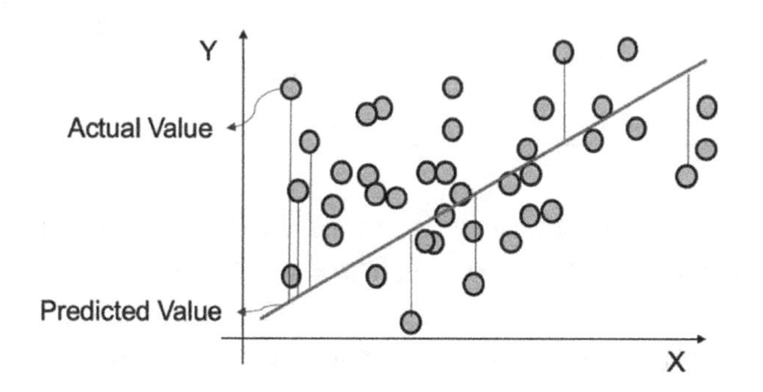

Figure 2-6. *The difference between the actual values and predicted values of the target variable. This is the error made while making the prediction, and as a best model, this error is to be minimized.*

A very important point: why do we take a square of the errors? The reason is that if we do not take the squares of the errors, the positive and negative terms can cancel each other. For example, if the $error_1$ is +2 and $error_2$ is -2, then net error will be 0!

From Figure 2-6, we can say that the

$$\begin{aligned}\text{Minimum squared sum of error} &= \min \Sigma e_i^2 \\ &= \min \Sigma\left(Y_i - \hat{Y}_i\right)^2 \qquad \text{(Equation 2-2)} \\ &= \min \Sigma\left[Y_i - (\beta_0 + \beta_1 x_i)\right]^2\end{aligned}$$

The estimated slope coefficient is

$$\beta_1 = \frac{\sum_{i=1}^{n}(x_i - \bar{x})(y_i - \bar{y})}{\sum_{i=1}^{n}(x_i - \bar{x})^2} \qquad \text{(Equation 2-3)}$$

The estimated intercept coefficient is

$$\beta_0 = \bar{y} - b_1 \bar{x} \qquad \text{(Equation 2-4)}$$

A point to be noted is that the regression line always passes through the mean $\bar{x}$ and $\bar{y}$.

Based on the preceding discussion, we can study measures of variation. The total variation is made of two parts, which can be represented by the following equation:

SST	=	SSR	+	SSE
Total sum of squares	=	Regression sum of squares	+	Error sum of squares
$SST = \Sigma(y_i - \bar{y})^2$		$SSR = \Sigma(\hat{Y}_i - \bar{y})^2$		$SSE = \Sigma(y_i - \hat{Y}_i)^2$

Where $\bar{y}$: Average value of dependent variable

Y_i: Observed value of the dependent variable

$\hat{Y}_i$: Predicted value of y for the given value of xi

In the preceding equation, SST is the total sum of squares. It measures the variation of y_i around their mean $\bar{y}$. SSR is a regression sum of squares, which is the explained variation attributable to the linear relationship between x and y. SSE is the error sum of squares, which is the variation attributable to factors other than the linear relationship between x and y. The best way to understand this concept is by means of Figure 2-7.

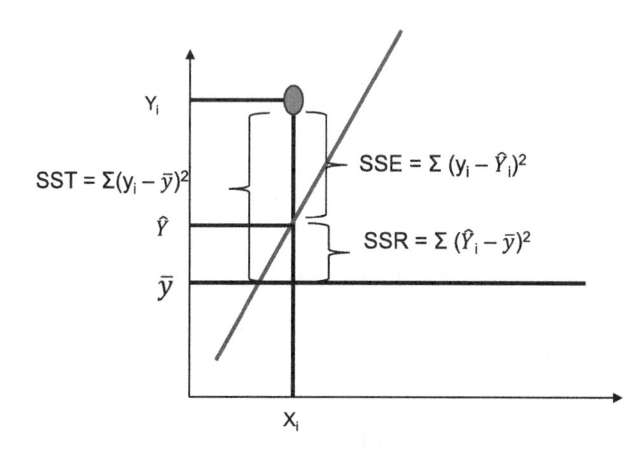

Figure 2-7. *SST = Total sum of squares: SSE = Error sum of squares, and SSR = Regression sum of squares*

The concepts discussed in the preceding frame a foundation for us to study the various measures and parameters to check the accuracy of a regression model, which we are discussing now:

1. Mean absolute error (MAE): As the name suggests, it is the average of the absolute difference between the actual and predicted values of a target variable as shown in Equation 2-5.

$$\text{Mean Absolute error} = \frac{\Sigma\left(\left|\hat{Y}_i - y_i\right|\right)}{n} \qquad \text{(Equation 2-5)}$$

The greater the value of MAE, the greater the error in our model.

2. Mean squared error (MSE): MSE is the average of the square of the error, that is, the difference between the actual and predicted values as shown in Equation 2-6. Similar to MAE, a higher value of MSE means higher error in our model.

$$\text{Mean squared error} = \frac{\Sigma\left(\left|\hat{Y}_i - y_i\right|\right)^2}{n} \quad \text{(Equation 2-6)}$$

3. Root MSE: Root MSE is the square root of the average squared error and is represented by Equation 2-7.

$$\text{Root mean squared error} = \sqrt{\frac{\Sigma\left(\left|\hat{Y}_i - y_i\right|\right)^2}{n}} \quad \text{(Equation 2-7)}$$

4. R square (R^2): It represents how much randomness in the data is being explained by our model. In other words, out of the total variation in our data how much is our model able to decipher.

R^2 = SSR/SST = Regression sum of squares/Total sum or squares

R^2 will always be between 0 and 1 or 0% and 100%. The higher the value of R^2, the better it is. The way to visualize R^2 is depicted in the following figures. In Figure 2-8, the value of R^2 is equal to 1, which means 100% of the variation in the value of Y is explainable by x. In Figure 2-9, the value of R^2 is between 0 and 1, depicting that some of the variation is understood by the model. In the last case, Figure 2-9, $R^2 = 0$, which shows that no variation is understood by the model. In a normal business scenario, we get R^2 between 0 and 1, that is, a portion of the variation is explained by the model.

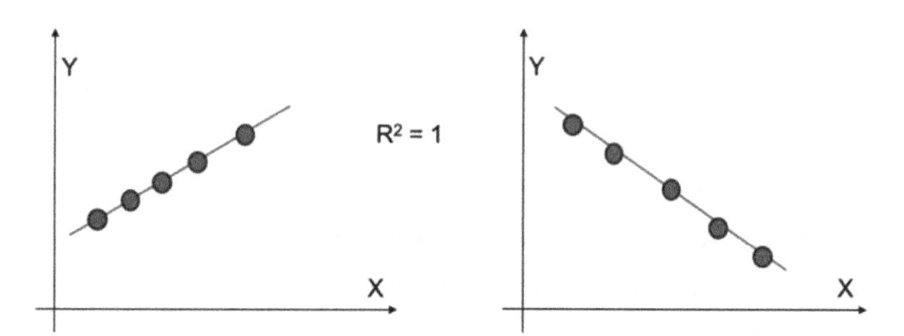

Figure 2-8. *R^2 is equal to 1; this shows that 100% of the variation in the values of the independent variable is explained by our regression model*

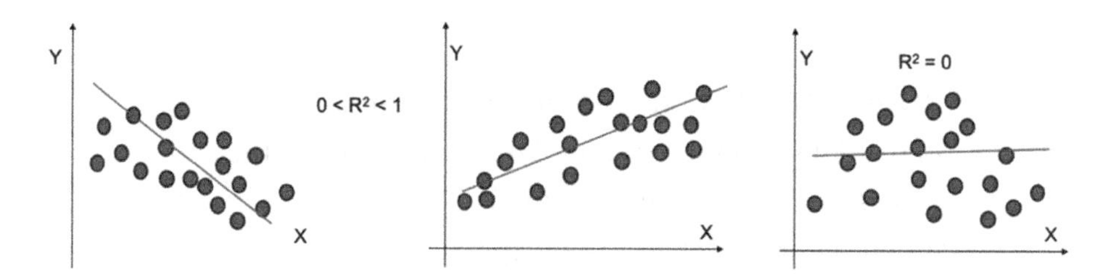

Figure 2-9. *If the value of R^2 is between 0 and 1 then partial variation is explained by the model. If the value is 0, then no variation is explained by the regression model*

5. Pseudo R square: It extends the concept of R square. It penalizes the value if we include insignificant variables in the model. We can calculate pseudo R^2 as in Equation 2-8.

$$\text{Pseudo R}^2 = \frac{\dfrac{1-\text{SSE}}{(n-k-1)}}{\dfrac{\text{SST}}{(n-1)}} \qquad \text{(Equation 2-8)}$$

where n is the sample size and k is the number of independent variables

Using R square, we measure all the randomness explained in the model. But if we include all the independent variables including insignificant ones, we cannot claim that the model is robust. Hence, we can say pseudo R square is a better representation and measure to measure the robustness of a model.

Tip Between R^2 and pseudo R^2, we prefer pseudo R^2. The higher the value, the better the model.

Now we are clear on the assumptions of a regression model and how we can measure the performance of a regression model. Now let us study types of linear regression and then develop a Python solution.

Linear regression can be studied in two formats: simple linear regression and multiple linear regression.

Simple linear regression is, as its name implies, simple to understand and easy to use. It is used to predict the value of the target variable using only one independent variable. As in the preceding example, if we have only the house area as an independent variable to predict the house prices, it is an example of simple linear regression. The sample data have only the area and the price is shown in the following. In Figure 2-10, we have the scatter plot of the same data.

Square feet (x)	Price (in 1000 $)
1200	100
1500	120
1600	125

(*continued*)

Square feet (x)	Price (in 1000 $)
1100	95
1500	125
2200	200
2100	195
1410	110

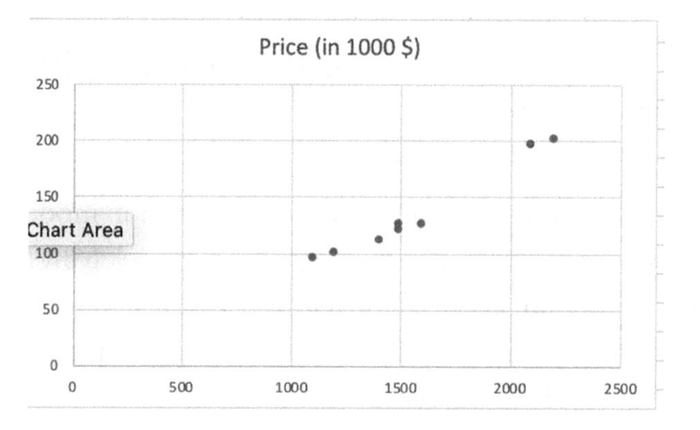

***Figure 2-10.** Scatter plot of the price and area data*

We can represent the data in Equation 2-9.

$$\text{Price} = \beta_0 + \beta_1 * \text{Area} \qquad \text{(Equation 2-9)}$$

Here price is Y (dependent variable) and area is the x variable (independent variable). The goal of the simple linear regression will be to estimate the values of β_0 and β_1 so that we can predict the values of price.

If we run a regression analysis, we will get the respective values of coefficients β_0 and β_1 as –28.07 and 0.10. We will be running the code using Python later, but let's interpret the meaning for understanding.

1. β_0 having a value of –28.07 means that when the square footage is 0 then the price is –$28,070. Now it is not possible to have a house with 0 area; it just indicates that for houses within the range of sizes observed, $28,070 is the portion of the house price not explained by the area.

2. β_1 is 0.10;it indicates that the average value of a house increases by a $0.1 \times (1000) =$ $100 with on average for each additional increase in the size by one square foot.

There is an important limitation to linear regression. We cannot use it to make predictions beyond the limit of the variables we have used in training. For example, in the preceding data set we cannot model to predict the house prices for areas above 2200 square feet and below 1100 square feet. This is because we have not trained the model for such values. Hence, the model is good enough for the values between the minimum and maximum limits only; in the preceding example the valid range is 1100 square feet to 2200 square feet.

It is time to hit the lab and solve the problem using a dataset in Python. We are using Python and the following are a few of the important attributes in a Python code. These are used when we train the data, set the parameters, and make a prediction:

- Common hyperparameters
    - fit_interceprt: if we want to calculate intercept for the model, and it is not required if data is centered
    - normalize - X will be normalized by subtracting mean & dividing by standard deviation
    - By standardizing data before subjecting to model, coefficients tell the importance of features

- Common attributes
    - coefficients are the weights for each independent variable
    - intercept is the bias of independent term of linear models
- Common functions
    - fit - trains the model. Takes X & Y as the input value and trains the model
    - predict - once model is trained, for given X using predict function Y can be predicted
- Training model
    - X should be in rows of data format, X.ndim == 2
    - Y should be 1D for single target & 2D for more than one target
    - fit function is used for training the model

We will be creating two examples for simple linear regression. In the first example, we are going to generate a simple linear regression problem. Then in example 2, simple linear regression problem will be solved. Let us proceed to example 1 now.

Example 1: Creating a Simple Linear Regression

We are going to create a simple linear regression by generating a dataset. This code snippet is only a warm-up exercise and will familiarize you with the simplest form of simple linear regression.

Step 1: Import the necessary libraries in the first step. We are going to import numpy, pandas, matplotlib, and scikit learn.

```
import numpy as np
import pandas as pd
import matplotlib.pyplot as plt
%matplotlib inline
from sklearn.linear_model import LinearRegression
from sklearn.datasets import make_regression
```

Step 2: In the next step, we are going to create a sample dataset which can be used for regression.

```
X,Y = make_regression(n_features=1, noise=5, n_samples=5000)
```

In the preceding code, `n_features` is the number of features we want to have in the dataset. n_samples is the number of samples we want to generate. Noise is the standard deviation of the gaussian noise which gets applied to the output.

Step 3: Plot the data using matplotlib library. Xlabel and ylabel will give the labels to the figure.

```
plt.xlabel('Feature - X')
plt.ylabel('Target - Y')
plt.scatter(X,Y,s=5)
```

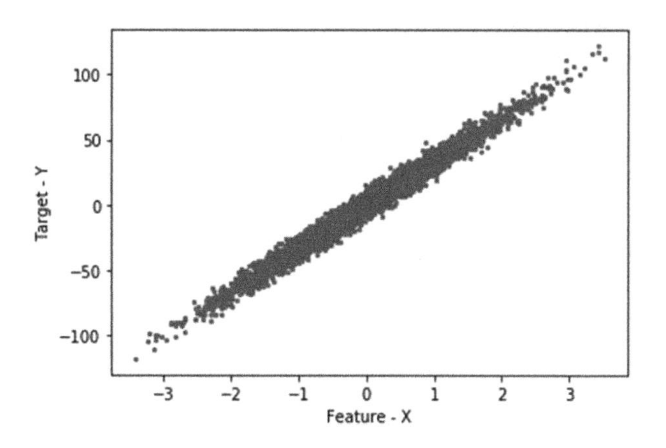

Step 4: Initialize an instance of linear regression now. The name of the variable is linear_model

```
linear_model = LinearRegression()
```

Step 5: Fit the linear regression now. The input independent variable is X and target variable is Y

```
linear_model.fit(X,Y)
```

Step 6: The model is trained now. Let us have a look at the coefficient values for both the intercept and the slope of the linear regression model

```
linear_model.coef_
linear_model.intercept_
```

The value for the intercept is 0.6759344 and slope is 33.1810

Step 7: We use this trained model to predict the value using X and then plot it

```
pred = linear_model.predict(X)
plt.scatter(X,Y,s=25, label='training')
plt.scatter(X,pred,s=25, label='prediction')
plt.xlabel('Feature - X')
plt.ylabel('Target - Y')
plt.legend()
plt.show()
```

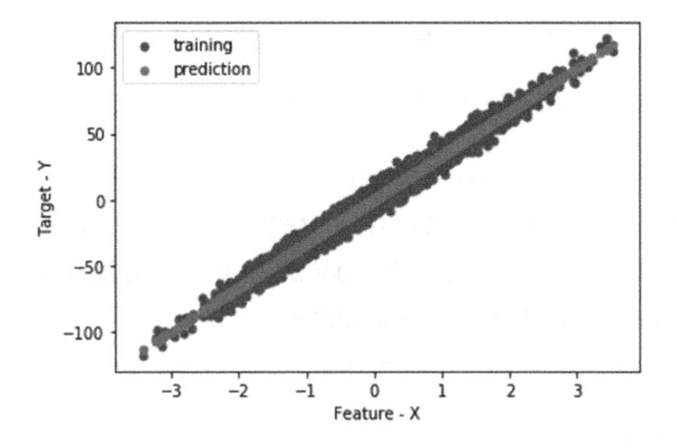

Blue dots represent maps to actual target data while orange dots represent the predicted values.

We can see that how close the training and actual predicted values are. Now, we will create a solution for simple linear regression using a dataset.

Example 2: Simple Linear Regression for Housing Dataset

We have a dataset which has one independent variable (area in square feet), and we have to predict the prices. It is again an example of simple linear regression, that is, we have one input variable. The code and dataset are uploaded to the Github link shared at the start of this chapter.

Step 1: Import all the required libraries here:

```
import pandas as pd
import numpy as np
import matplotlib.pyplot as plt
import seaborn as sns
%matplotlib inline
import warnings
warnings.filterwarnings(action="ignore", module="scipy",
message="^internal gelsd")
```

Step 2: Load the data set using pandas function:

```
house_df= pd.read_csv('House_data_LR.csv')
```

Note It is a possibility that data is present in the form of .xls or .txt file. Sometimes, the data is to be loaded from the database directly by making a connection to the database.

```
house_df.head()
```

	Unnamed: 0	sqft_living	price
0	0	1180	221900.0
1	1	2570	538000.0
2	2	770	180000.0
3	3	1960	604000.0
4	4	1680	510000.0

Step 3: Check if there is any null value present in the dataset:

```
house_df.isnull().any()
```

Step 4: There is a variable which does not make sense. We are dropping the variable "Unnamed":

```
house_df.drop('Unnamed: 0', axis = 1, inplace = True)
```

Step 5: After dropping the variable, let's have a look at the first few rows in the dataset:

```
house_df.head()
```

	sqft_living	price
0	1180	221900.0
1	2570	538000.0
2	770	180000.0
3	1960	604000.0
4	1680	510000.0

Step 6: We will now prepare the dataset for model building by separating the independent variable and target variable.

```
X = house_df.iloc[:, :1].values
y = house_df.iloc[:, -1].values
```

Step 7: The data is split into train and test now.

Train/Test Split: Creating a train and test dataset involves splitting the dataset into training and testing sets respectively, which are mutually exclusive. After which, you train with the training set and test with the testing set. This will provide a more accurate evaluation of out-of-sample accuracy because the testing dataset is not part of the dataset that has been used to train the data. It is more realistic for real-world problems.

This means that we know the outcome of each data point in this dataset, making it great to test with! And since this data has not been used to train the model, the model has no knowledge of the outcome of these data points. So, in essence, it is truly an out-of-sample testing. Here test data is 25% or 0.25.

Random state, as the name suggests, is for initializing the internal random number generator, which in turn decides the splitting the train/test indices. Keeping it fixed allows us to replicate the same train/test split and hence in verification of the output.

```
from sklearn.model_selection import train_test_split
X_train, X_test, y_train, y_test = train_test_split(X, y,
test_size = 0.25, random_state = 5)
```

Step 8: Fit the data now using the linear regression model:

```
from sklearn.linear_model import LinearRegression
simple_lr= LinearRegression()
simple_lr.fit(X_train, y_train)
```

Step 9: The model is trained now. Let us use it to predict on the test data

```
y_pred = simple_lr.predict(X_test)
```

Step 10: We will first test the model on the training data. We will try to predict on training data and visualize the results on it.

```
plt.scatter(X_train, y_train, color = 'r')
plt.plot(X_train, simple_lr.predict(X_train), color = 'b')
plt.title('Sqft Living vs Price for Training')
plt.xlabel('Square feet')
plt.ylabel('House Price')
plt.show()
```

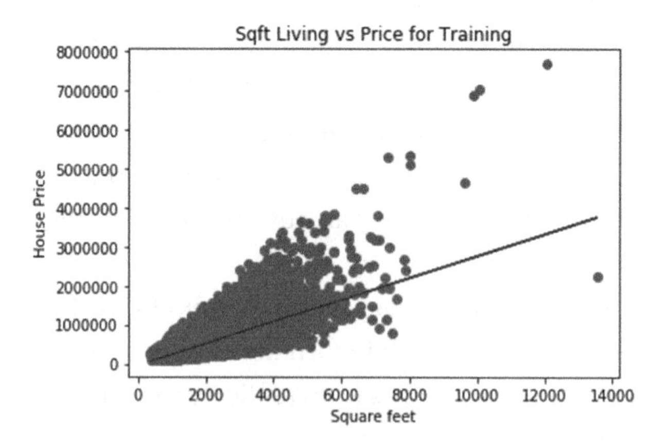

Step 11: Now let us test the model on the testing data. It is the correct measurement to check how robust the model is.

```
plt.scatter(X_test, y_test, color = 'r')
plt.plot(X_train, simple_lr.predict(X_train), color = 'b')
plt.title('Sqft Living vs Price for Test')
plt.xlabel('Square feet')
plt.ylabel('House Price')
```

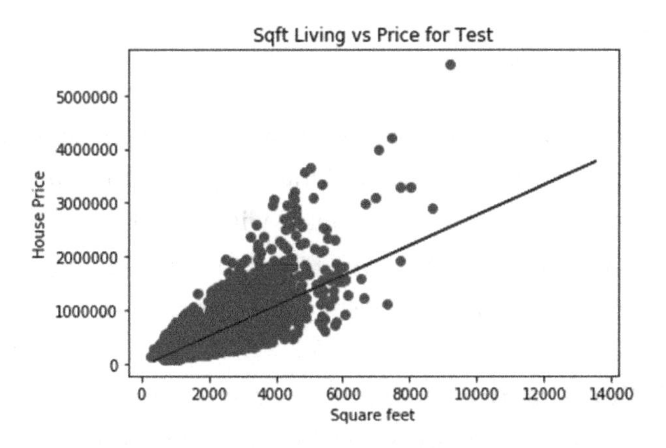

Step 12: Now let's figure out how good or how bad we are doing in the predictions. We will calculate the MSE and R^2.

```
from sklearn.metrics import mean_squared_error
from math import sqrt
rmse = sqrt(mean_squared_error(y_test, y_pred))
from sklearn.metrics import r2_score
r2 = r2_score(y_test, y_pred)
adj_r2 = 1 - float(len(y)-1)/(len(y)-len(simple_lr.coef_)-1)*(1 - r2)
rmse, r2, adj_r2, simple_lr.coef_, simple_lr.intercept_
```

```
(257125.13804007217,
 0.5020612063135523,
 0.5020381653254589,
 array([281.4054356]),
 -45441.30813530844)
```

Step 13: We will now make a prediction on unseen value of x:

```
import numpy as np
x_unseen=np.array([1500]).reshape(1,1)
simple_lr.predict(x_unseen)
The prediction is 376666.84
```

In the preceding two examples, we saw how we can use simple linear regression to train a model and make a prediction. In real-world problems, only one independent variable will almost never happen. Most business world problems have more than one variable, and such problems are solved using multiple linear regression algorithms which we are discussing now.

Multiple linear regression or multiple regression can be said as an extension of simple linear regression where instead of one independent variable we have more than one independent variable. For example, Figure 2-11 shows the representation of more than one variable of a similar dataset in the vector-space diagram:

Square Feet	No of bedrooms	Price (in 1000 $)
1200	2	100
1500	3	120
1600	3	125
1100	2	95
1500	2	125

(continued)

Square Feet	No of bedrooms	Price (in 1000 $)
2200	4	200
2100	4	195
1410	2	110

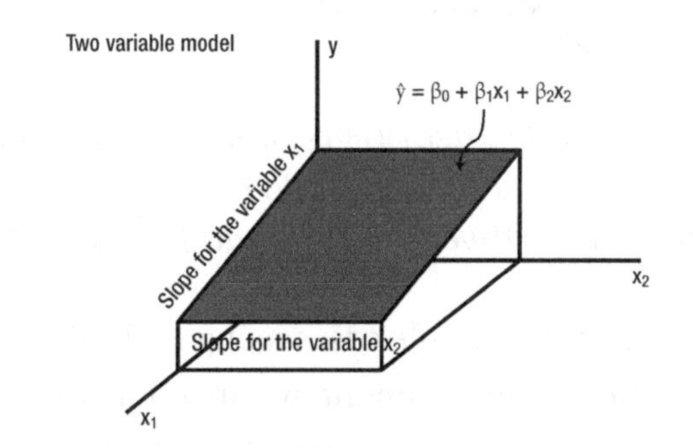

***Figure 2-11.** Multiple regression model depiction in vector-space diagram where we have two independent variables, x_1 and x_2*

Hence the equation for multiple linear regression is shown in Equation 2-10.

$$Y_i = \beta_0 + \beta_1 x_1 + \beta_2 x_2 + \ldots + \varepsilon_i \qquad \text{(Equation 2-10)}$$

Hence in the case of a multiple training of the simple linear regression, we will get an estimated value for β_0 and β_1 and so on.

The residuals in the case of a multiple regression model are shown in Figure 2-12. The example is for a two-variable model, and we can clearly visualize the value of residual, which is the difference between actual and predicted values.

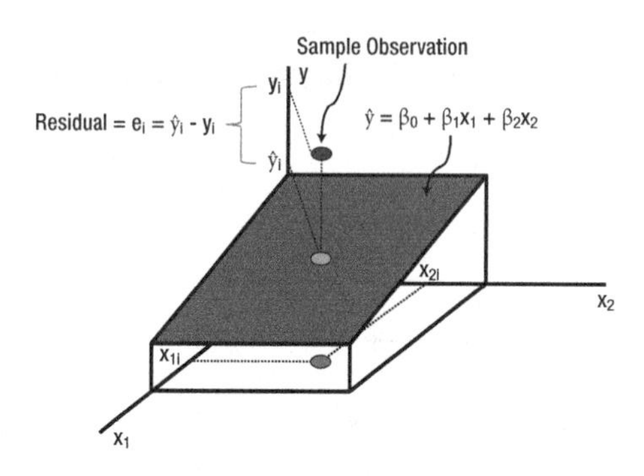

Figure 2-12. *The multiple linear regression model depicted by x_1 and x_2 and the residual shown with respect to a sample observation. The residual is the difference between actual and predicted values.*

We will now create two example cases using multiple linear regression models. During the model development, we are going to do EDA, which is the first step and will also resolve the problem of null values in the data and how to handle categorical variables too.

Example 3: Multiple Linear Regression for Housing Dataset

We are working on the house price dataset. The target variable is the prediction of house prices and there are some independent variables. The dataset and the codes are uploaded to the Github link shared at the start of this chapter.

Step 1: Import all the required libraries first.

```
import pandas as pd
import numpy as np
import matplotlib.pyplot as plt
import seaborn as sns
%matplotlib inline
```

```
import warnings
warnings.filterwarnings(action="ignore", module="scipy",
message="^internal gelsd")
```

Step 2: Import the data which is in the form of .csv file. Then check the first few rows.

```
 house_mlr = pd.read_csv('House_data.csv')
house_mlr.head()
```

Out[11]:

	id	date	price	bedrooms	bathrooms	sqft_living	sqft_lot	floors	waterfront	view	...	grade	sqft_above	sqft_basement	yr_built
0	7129300520	20141013T000000	221900.0	3	1.00	1180	5650	1.0	0	0	...	7	1180	0	1955
1	6414100192	20141209T000000	538000.0	3	2.25	2570	7242	2.0	0	0	...	7	2170	400	1951
2	5631500400	20150225T000000	180000.0	2	1.00	770	10000	1.0	0	0	...	6	770	0	1933
3	2487200875	20141209T000000	604000.0	4	3.00	1960	5000	1.0	0	0	...	7	1050	910	1965
4	1954400510	20150218T000000	510000.0	3	2.00	1680	8080	1.0	0	0	...	8	1680	0	1987

5 rows × 21 columns

We have 21 variables in this dataset.

Step 3: Next let's explore the dataset we have. This is done using the house_mlr.info() command.

```
<class 'pandas.core.frame.DataFrame'>
RangeIndex: 21613 entries, 0 to 21612
Data columns (total 21 columns):
id               21613 non-null int64
date             21613 non-null object
price            21613 non-null float64
bedrooms         21613 non-null int64
bathrooms        21613 non-null float64
sqft_living      21613 non-null int64
sqft_lot         21613 non-null int64
floors           21613 non-null float64
waterfront       21613 non-null int64
view             21613 non-null int64
condition        21613 non-null int64
grade            21613 non-null int64
sqft_above       21613 non-null int64
sqft_basement    21613 non-null int64
yr_built         21613 non-null int64
yr_renovated     21613 non-null int64
zipcode          21613 non-null int64
lat              21613 non-null float64
long             21613 non-null float64
sqft_living15    21613 non-null int64
sqft_lot15       21613 non-null int64
dtypes: float64(5), int64(15), object(1)
memory usage: 3.5+ MB
```

By analyzing the output we can see that out of the 21 variables, there are a few float variables, some object and some integers. We will be treating these categorical variables to integer variables.

Step 4: house_mlr.describe() command will give the details about all the numeric variables.

	id	price	bedrooms	bathrooms	sqft_living	sqft_lot	floors	waterfront	view
count	2.161300e+04	2.161300e+04	21613.000000	21613.000000	21613.000000	2.161300e+04	21613.000000	21613.000000	21613.000000
mean	4.580302e+09	5.400881e+05	3.370842	2.114757	2079.899736	1.510697e+04	1.494309	0.007542	0.234303
std	2.876566e+09	3.671272e+05	0.930062	0.770163	918.440897	4.142051e+04	0.539989	0.086517	0.766318
min	1.000102e+06	7.500000e+04	0.000000	0.000000	290.000000	5.200000e+02	1.000000	0.000000	0.000000
25%	2.123049e+09	3.219500e+05	3.000000	1.750000	1427.000000	5.040000e+03	1.000000	0.000000	0.000000
50%	3.904930e+09	4.500000e+05	3.000000	2.250000	1910.000000	7.618000e+03	1.500000	0.000000	0.000000
75%	7.308900e+09	6.450000e+05	4.000000	2.500000	2550.000000	1.068800e+04	2.000000	0.000000	0.000000
max	9.900000e+09	7.700000e+06	33.000000	8.000000	13540.000000	1.651359e+06	3.500000	1.000000	4.000000

Here we can see the range for all the numeric variables: the mean, standard deviation, the values at the 25th percentile, 50th percentile, and 75th percentile. The minimum and maximum values are also shown.

A very good way to visualize the variables is using a box-and-whisker plot using the following code.

```
fig = plt.figure(1, figsize=(9, 6))
ax = fig.add_subplot(111)
ax.boxplot(house_mlr['sqft_living15'])
```

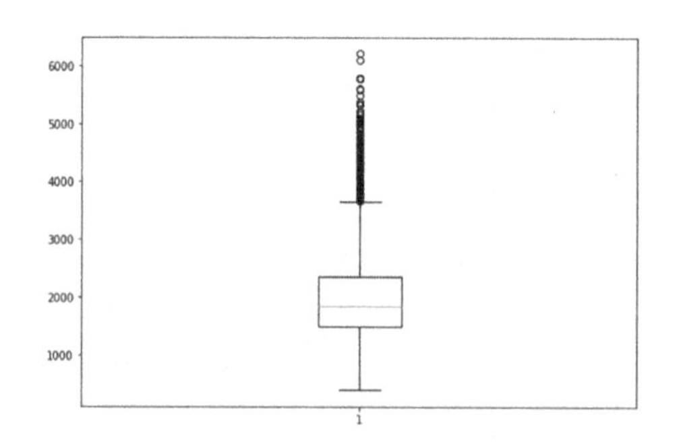

The plot shows that there are a few outliers. In this case, we are not treating the outliers. In later chapters, we shall examine the best practices to deal with outliers.

Step 5: Now we are going to check for the correlations between the variables. This will be done using a correlation matrix which is developed using the following code:

```
house_mlr.drop(['id', 'date'], axis = 1, inplace = True)
fig, ax = plt.subplots(figsize = (12,12))
ax = sns.heatmap(house_mlr.corr(),annot = True)
```

The analysis of the correlation matrix shows that there is some correlation between a few variables. For example, between sqft_above and sqft_living there is a correlation of 0.88. And that is quite expected. For this first simple example, we are not treating the correlated variables.

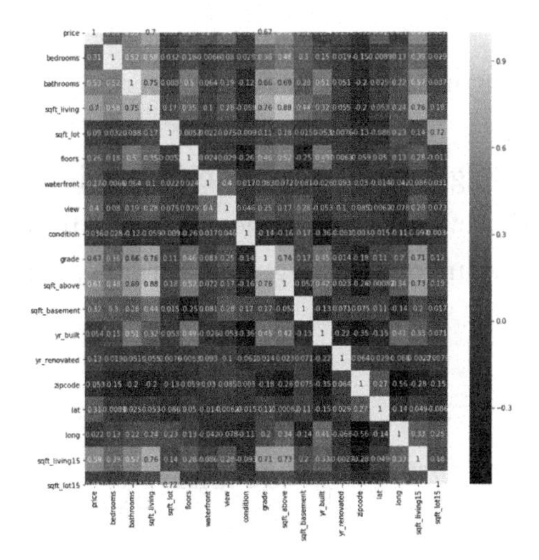

Step 6: Now we will clean the data a little. There are a few null values present in the dataset. We are dropping those null values now. We are examining the concepts of missing value treatment in Chapter 5.

```
house_mlr.isnull().any()
house_mlr ['basement'] = (house_mlr ['sqft_basement'] > 0).
astype(int)
```

```
house_mlr ['renovated'] = (house_mlr ['yr_renovated'] > 0).
astype(int)

to_drop = ['sqft_basement', 'yr_renovated']
house_mlr.drop(to_drop, axis = 1, inplace = True)

house_mlr.head()
```

	price	bedrooms	bathrooms	sqft_living	sqft_lot	floors	waterfront	view	condition	grade	sqft_above	yr_built	zipcode	lat
0	221900.0	3	1.00	1180	5650	1.0	0	0	3	7	1180	1955	98178	47.5112
1	538000.0	3	2.25	2570	7242	2.0	0	0	3	7	2170	1951	98125	47.7210
2	180000.0	2	1.00	770	10000	1.0	0	0	3	6	770	1933	98028	47.7379
3	604000.0	4	3.00	1960	5000	1.0	0	0	5	7	1050	1965	98136	47.5208
4	510000.0	3	2.00	1680	8080	1.0	0	0	3	8	1680	1987	98074	47.6168

Step 7: The categorical variables are converted to numeric ones using one-hot encoding.

One-hot encoding converts categorical variables to numeric ones. Simply put, it adds new columns to the dataset with 0 or assigned depending on the value of the categorical variable, as shown in the following:

CustID	Revenue	City	Items
1001	100	New Delhi	4
1002	101	London	5
1003	102	Tokyo	6
1004	104	New Delhi	8
1001	100	New York	4
1005	105	London	5

→

CustID	Revenue	New Delhi	London	Tokyo	New York	Items
1001	100	1	0	0	0	4
1002	101	0	1	0	0	5
1003	102	0	0	1	0	6
1004	104	1	0	0	0	8
1001	100	0	0	0	1	4
1005	105	0	1	0	0	5

```
categorical_variables = ['waterfront', 'view', 'condition',
'grade', 'floors','zipcode']

house_mlr = pd.get_dummies(house_mlr, columns = categorical_
variables, drop_first=True)
house_mlr.head()
```

	price	bedrooms	bathrooms	sqft_living	sqft_lot	sqft_above	yr_built	lat	long	sqft_living15	...	zipcode_98146	zipcode_98148
0	221900.0	3	1.00	1180	5650	1180	1955	47.5112	-122.257	1340	...	0	0
1	538000.0	3	2.25	2570	7242	2170	1951	47.7210	-122.319	1690	...	0	0
2	180000.0	2	1.00	770	10000	770	1933	47.7379	-122.233	2720	...	0	0
3	604000.0	4	3.00	1960	5000	1050	1965	47.5208	-122.393	1360	...	0	0
4	510000.0	3	2.00	1680	8080	1680	1987	47.6168	-122.045	1800	...	0	0

5 rows × 107 columns

Step 8: We will now split the data into train and test and then fit the model. Test size is 25% of the data.

```
X = house_mlr.iloc[:, 1:].values
y = house_mlr.iloc[:, 0].values

from sklearn.model_selection import train_test_split
X_train, X_test, y_train, y_test = train_test_split(X, y,
test_size = 0.25, random_state = 5)

from sklearn.linear_model import LinearRegression
multiple_regression = LinearRegression()
multiple_regression.fit(X_train, y_train)
```

Step 9: Predict the test set results.

```
y_pred = multiple_regression.predict(X_test)
```

Step 10: We will now check the accuracy of our model.

```
from sklearn.metrics import mean_squared_error
from math import sqrt
rmse = sqrt(mean_squared_error(y_test, y_pred))

from sklearn.metrics import r2_score
r2 = r2_score(y_test, y_pred)

adj_r2 = 1 - float(len(y)-1)/(len(y)-len(multiple_regression.
coef_)-1)*(1 - r2)

rmse, r2, adj_r2
```

The output we receive is (147274.98522602883, 0.8366403242154088, 0.8358351477235848).

The steps used in this example can be extended to any example where we want to predict a continuous variable. In this problem, we have predicted the value of a continuous variable but we have *not* selected the significant variables from the list of available variables. Significant variables are the ones which are more important than other independent variables in making the predictions. There are multiple ways to shortlist the variables. We will discuss one of them using the ensemble technique in the next section. The popular methodology of using p-value is discussed in Chapter 3.

With this, we have discussed the concepts and implementation using linear regression. So far, we have assumed that the relationship between dependent and independent variables is linear. But what if this relation is not linear? That is our next topic: nonlinear regression.

Nonlinear Regression Analysis

Consider this. In physics, we have laws of motion to describe the relationship between a body and all the forces acting upon it and how the motion responds to those forces. In one of the laws, the relationship of initial and final velocity is given by the following equation:

$$\text{Final velocity} = \text{initial velocity} + \tfrac{1}{2}\,\text{acceleration}*\text{time}^2 \text{ OR } v = u + \tfrac{1}{2} at^2$$

If we analyze this equation, we can see that the relation between final velocity and time is not linear but quadratic in nature. *Nonlinear* regression is used to model such relationships. We can review the scatter plot to identify if the relationship is nonlinear as shown in Figure 2-13.

Formally put, if the relationship between target variable and independent variables is not linear in nature, then *nonlinear regression* is used to fit the data.

The model form for nonlinear regression is a mathematical equation, which can be represented as Equation 2-11 and as the curve in Figure 2-13. The shape of the curve will depend on the value of "n" and the respective values of β_0, β_1..., and so on.

$$Y_i = \beta_0 + \beta_1 x_1 + \beta_2 x^2_1 + \beta_n x^n_1 + \ldots + \varepsilon_i \quad \text{(Equation 2-11)}$$

Where

β_0: Y intercept

β_1: regression coefficient for linear effect of X on Y

β_2 = regression coefficient for quadratic effect on Y and so on

ε_i = random error in Y for observation i

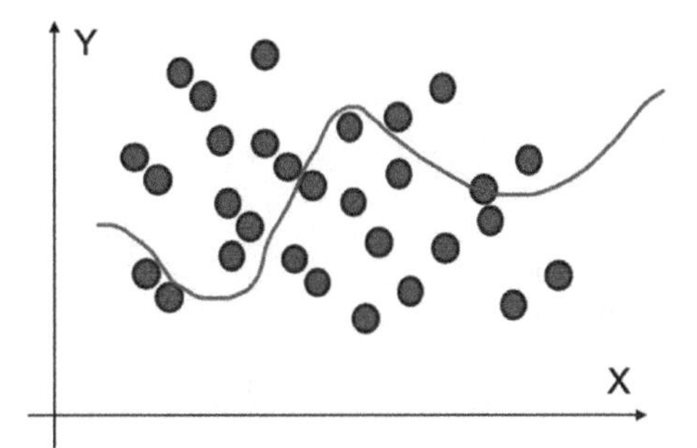

***Figure 2-13.** A nonlinear regression will do a better job to model the data set than a simple linear model. There is no linear relationship between the dependent and the independent variable*

Let's dive deeper and understand nonlinear relationships by taking quadratic equations as an example. If we have a quadratic relationship between the dependent variables, it can be represented by Equation 2-12.

$$Y_i = \beta_0 + \beta_1 x_{1i} + \beta_2 x^2_{1i} + \varepsilon_i \quad \text{(Equation 2-12)}$$

Where

β_0: Y intercept

β_1: regression coefficient for linear effect of X on Y

β_2 = regression coefficient for quadratic effect on Y

ε_i = random error in Y for observation i

The quadratic mode will take the following shapes depending on the values of β_1 and β_2 as shown in Figure 2-14, in which β_1 is the coefficient of the linear term and β_2 is the coefficient of the squared term.

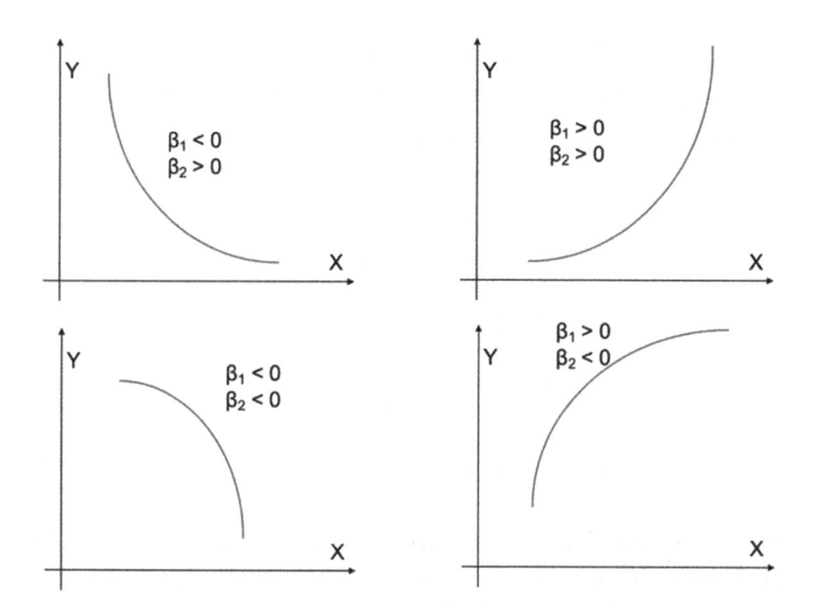

Figure 2-14. *The value of the curve is dependent on the values of β_1 and β_2. The figure depicts the different shapes of the curve in different directions.*

Depending on the values of the coefficients, the shape of the curve will change its shape. An example of such a case is shown in the following data. Figure 2-15 on the right side shows the relationship between two variables: velocity and distance:

Velocity	Distance
3	9
4	15
5	28
6	38
7	45
8	69
10	96
12	155
18	260
20	410
25	650

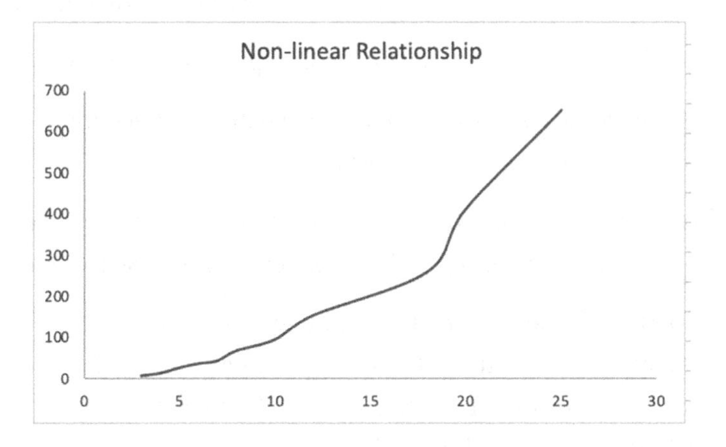

***Figure 2-15.** Nonlinear relationship between velocity and distance is depicted here*

The data and the respective plot of the data indicate nonlinear relationships. Still, we should know how to detect if a relationship between target and independent variables is nonlinear, which we are discussing next.

Identifying a Nonlinear Relationship

While we try to fit a nonlinear model, first we have to make sure that a nonlinear equation is indeed required. Testing for significance for the quadratic effect can be checked by the standard NULL hypothesis test.

1. The estimate in the case of linear regression is $\hat{y} = b_0 + b_1x_1$
2. The estimate in the case of quadratic regression is $\hat{y} = b_0 + b_1x_1 + b_2x^2_1$
3. The NULL hypothesis will be
    a. $H_0 : \beta_2 = 0$ (the quadratic term does not improve the model)
    b. $H_1 : \beta_2 \neq 0$ (the quadratic term improves the model)
4. Once we run the statistical test, we will either accept or reject the NULL hypothesis

But this might not always be feasible to run a statistical test. While working on practical business problems, we can take these two steps:

1. To identify the nonlinear relationship, we can analyze the residuals after fitting a linear model. If we try to fit a linear relationship while the actual relationship is nonlinear, then the residuals will not be random but will have a pattern, as shown in Figure 2-16. If a nonlinear relationship is modeled instead, the residuals will be random in nature.

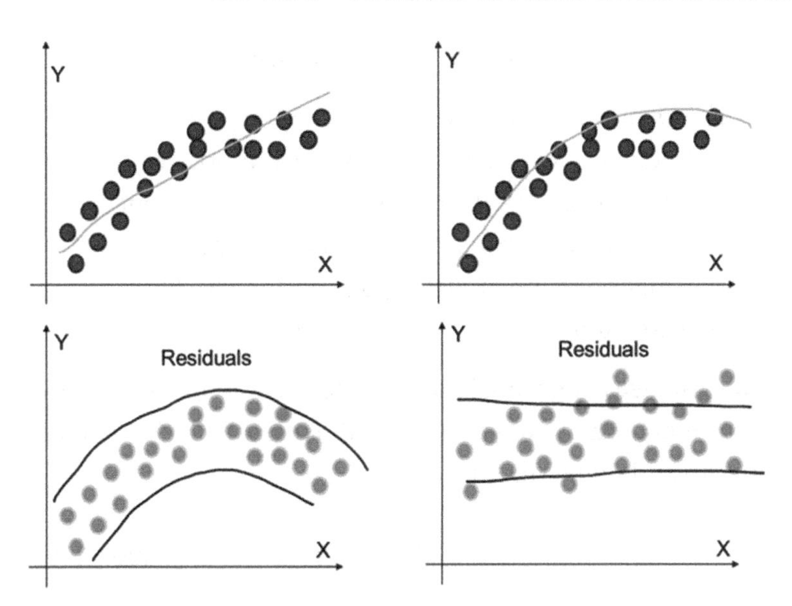

Figure 2-16. *If the residuals follow a pattern, it signifies that there can be a nonlinear relationship present in the data which is being modeled using a linear model. Linear fit does not give random residuals, while nonlinear fit will give random residuals.*

2. We can also compare the respective R^2 values of both the linear and nonlinear regression model. If the R^2 of the nonlinear model is more, it means the relationship with nonlinear model is better.

Similar to linear regression, nonlinear models too have some assumptions, which we will discuss now.

Assumptions for a Nonlinear Regression

1. The errors in a nonlinear regression follow a normal distribution. Error is the difference between the actual and predicted values and nonlinear requires the variables to follow a normal distribution.

2. All the errors must have a constant variance.
3. The errors are independent of each other and do not follow a pattern. This is a very important assumption since if the errors are not independent of each other, it means that there is still some information in the model equation which we have not extracted.

We can also use a **log transformation** to solve some of the nonlinear models. The equation of a log transformation can be put forward as Equation 2-13 and Equation 2-14.

$$Y = \beta_0 X_1^{\beta 1} X_2^{\beta 2} \varepsilon \qquad \text{(Equation 2-13)}$$

and by taking a log transformation on both the sides:

$$\log(Y) = \log(\beta_0) + \beta_1 \log(X_1) + \beta_2 \log(X_2) + \log(\varepsilon) \qquad \text{(Equation 2-14)}$$

The coefficient of the independent variable can be interpreted as follows: a 1% change in the independent variable X_1 leads to an estimated β_1 percentage change in the average value of Y.

Note Sometimes, β_1 can also refer to elasticity of Y with respect to change in X_1.

We have studied different types of regression models. It is one of the most stable solutions, but like any other tool or solution, there are a few pitfalls with regression models too. Some of those in the form of important insights will be uncovered while doing EDA. These challenges have to be dealt with while we are doing the data preparation for our statistical modeling. We will now discuss these challenges.

Challenges with a Regression Model

Though regression is quite a robust model to use, there are a few challenges we face with the algorithm:

1. **Nonlinearities**: Real-world data points are much more complex and generally do not follow a linear relationship. Even if we have a very large number of data points, a linear method might prove too simplistic to create a robust model, as shown in Figure 2-17. A linear model will not be able to do a good job for the one on the left, while if we have a nonlinear model the equation fits better. For example, seldom we will find the price and size following a linear relationship.

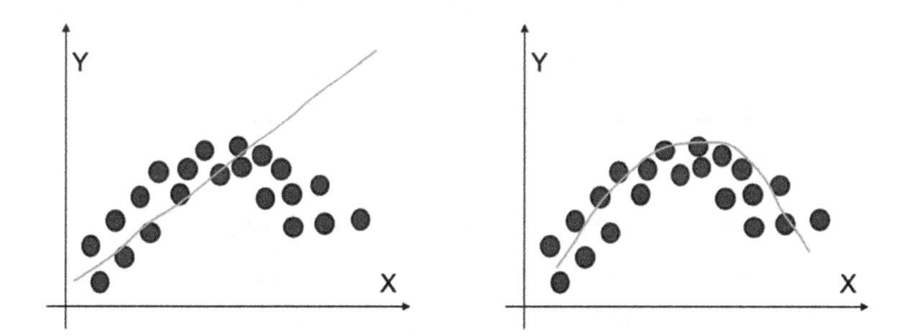

Figure 2-17. *The figure on the left shows that we are typing to model a linear model for nonlinear data. The figure on the right is the correct equation. Linear relationship is one of the important assumptions in linear regression*

2. **Multicollinearity**: We have discussed the concept of multicollinearity earlier in the chapter. If we use correlated variables in our model, we will face the problem of multicollinearity. For example, if we include as units both sales in thousands and as

a revenue in $, both the variables are essentially talking about the same information. If we have a problem of multicollinearity, it impacts the model as follows:

a. The estimated coefficients of the independent variables become quite sensitive to even small changes in the model and hence their values can change quickly.

b. The overall predictability power of our model takes a hit as the precision of the estimated coefficients of the independent variables can get impacted.

c. We may not be able to trust the p-value of the coefficients and hence we cannot completely rely on the significance shown by the model.

d. Hence, it undermines the overall quality of the trained model and this multicollinearity needs to be acted upon.

3. **Heteroscedasticity**: This is one more challenge we face while modeling a regression problem. The variability of the target variable is directly dependent on the change in the values of the independent variables. It creates a cone-shaped pattern in the residuals, which is visible in Figure 2-5. It creates a few challenges for us like the following:

    a. Heteroscedasticity messes with the significance of the independent variables. It inflates the variance of coefficient estimates. We would expect this increase to be detected by the OLS process but OLS is not able to. Hence, the t-values and F-values calculated are wrong and consequently the respective p-values estimated are less. That leads us to make incorrect conclusions about the significance of the independent variables.

b. Heteroscedasticity leads to incorrect coefficient estimates for the independent variables. And hence the resultant model will be biased.

4. **Outliers**: Outliers lead to a lot of problems in our regression model. It changes our results and makes a huge impact on our insights and the ML model. The impacts are as follows:

   a. Our model equation takes a serious hit in the case of outliers as visible in Figure 2-18. In the presence of outliers, the regression equation tries to fit them too and hence the actual equation will not be the best one.

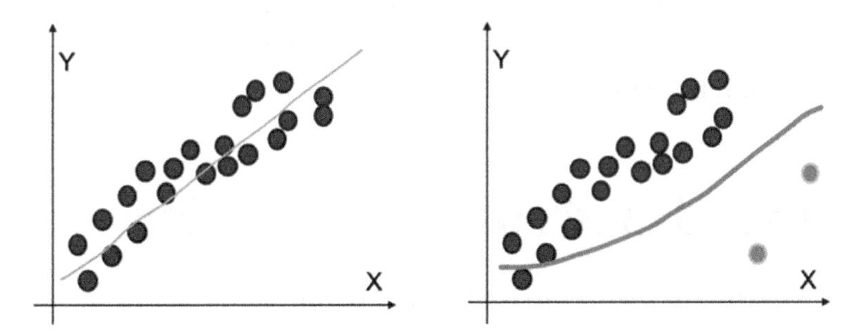

Figure 2-18. *Outliers in a dataset seriously impact the accuracy of the regression model since the equation will try to fit the outlier points too; hence, the results will be biased*

   b. Outliers bias the estimates for the model and increase the error variance.

c. If a statistical test is to be performed, its power and impact take a serious hit.

d. Overall, from the data analysis, we cannot trust coefficients of the model and hence all the insights from the model will be erroneous.

Note We will be dealing with how to detect outliers and how to tackle them in Chapter 5.

Linear regression is one of widely used techniques to predict the continuous variables. The usages are vast and many and across multiple domains. It is generally the first few methods we use to model a continuous variable and it acts as a benchmark for the other ML models.

This brings us to the end of our discussion on linear regression models. Next we will discuss a quite popular family of ML models called tree models or tree-based models. Tree-based models can be used for both regression and classification problems. In this chapter we will be studying only the regression problems and in the next chapter we will be working on classification problems.

Tree-Based Methods for Regression

The next type of algorithms used to solve ML problems are *tree-based algorithms*. Tree-based algorithms are very simple to comprehend and develop. They can be used for both regression and classification problems. A decision tree is a supervised learning algorithm, hence we have a target variable and depending on the problem it will be either a classification or a regression problem.

A decision tree looks like Figure 2-19. As you can see, the entire population is continuously split into groups and subgroups based on a criterion. We start with the entire population at the beginning of the tree and subsequently the population is divided into smaller subsets; at the same time, an associated decision tree is incrementally developed. Like we do in real life, we consider the most important factor first and then divide the possibilities around it. In a similar way, decision tree building starts by finding the features for the best splitting condition.

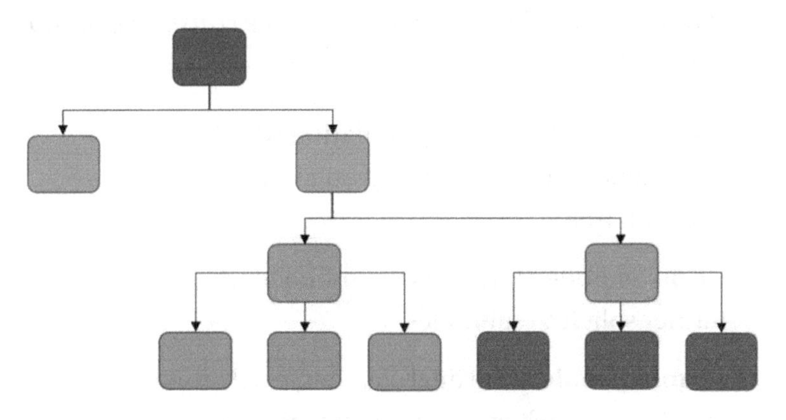

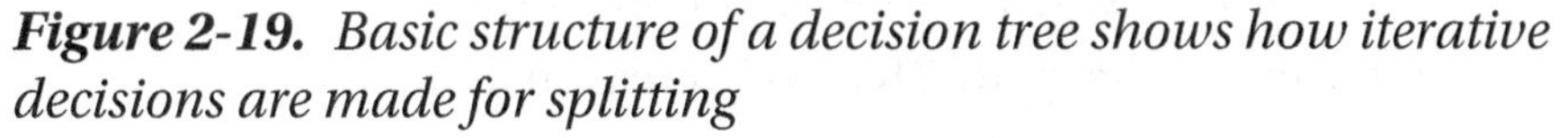

***Figure 2-19.** Basic structure of a decision tree shows how iterative decisions are made for splitting*

Decision trees can be used to predict both continuous variables and categorical variables. In the case of a regression tree, the value achieved by the terminal node is the mean of the values falling in that region. While in the case of classification trees, it is the mode of the observations. We will be discussing both the methods in this book. In this chapter, we are examining regression problems; in the next chapter classification problems are solved using decision trees.

Before moving ahead, it is imperative we study the building blocks for a decision tree. As you can see in Figure 2-20, a decision tree is represented using the root node, decision node, terminal node, and a branch.

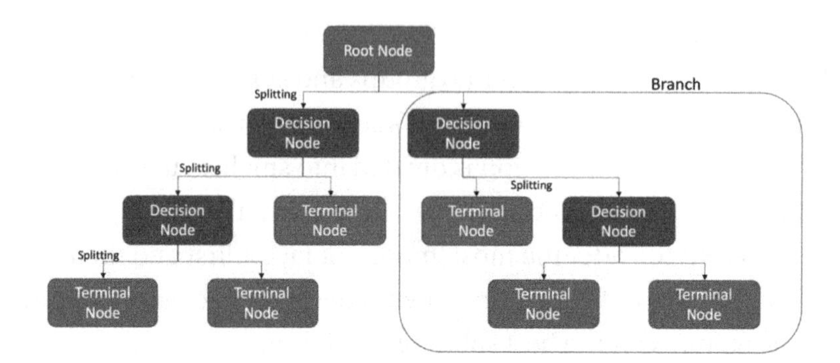

***Figure 2-20.** Building blocks of a decision tree consisting of root node, decision node, terminal node, and a branch*

1. **Root node** is the entire population which is being analyzed and is displayed at the top of the decision tree.

2. **Decision node** represents the subnode which gets further split into subnodes.

3. **Terminal node** is the final element in a decision tree. Basically when a node cannot split further, it is the end of that path. That node is called *terminal* node. Sometimes it is also referred to as *leaf.*

4. **Branch** is a subsection of a tree. It is sometimes called a *subtree.*

5. **Parent node and child node** are the references made to nodes only. A node which is divided is a parent node and the subnodes are called the child nodes.

Let us now understand the decision tree using the house price prediction problem. For the purpose of understanding, let us assume the first criteria of splitting is area. If the area is less than 100 sq km. then the entire population is split in two nodes as shown in Figure 2-21(i). On the right hand, we can see the next criteria is number of bedrooms. If the number of bedrooms is less than four, the predicted price is 100; otherwise, the predicted price is 150. For the left side the criteria for splitting is distance. And this process will continue to predict the price values.

A decision tree can also be thought as a group of nested IF-ELSE conditions (as shown in Figure 2-21(ii)) which can be modeled as a tree wherein the decisions are made in the internal node and the output is obtained in the leaf node. A decision tree divides the independent variables into non-overlapping spaces. These spaces are distinct to each other too. Geometrically, a decision tree can also be thought of as a set of parallel hyperplanes which divide the space into a number of hypercuboids as shown in Figure 2-21(iii). We predict the value for the unseen data based on which hypercuboid it falls into.

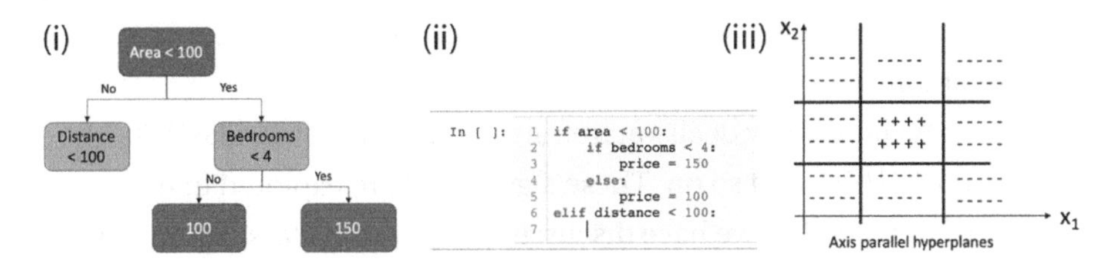

***Figure 2-21.** (i) Decision tree based split for the housing prediction problem. (ii) Decision tree can be thought as a nested IF-ELSE block. (iii) Geometric representation of a decision tree shows parallel hyperplanes.*

Now you have understood what a decision tree is. Let us also examine the criteria of splitting a node.

A decision tree utilizes a *top-down greedy approach.* As we know, a decision tree starts with the entire population and then recursively splits the data; hence it is called *top-down.* It is called a greedy approach, as the algorithm at the time of decision of split takes the decision for the current split only based on the best available criteria, that is, variable and not based on the future splits, which may result in a better model. In other words, for greedy approaches the focus is on the current split only and not the future splits. This splitting takes place recursively unless the tree is fully grown and the stopping criteria is reached. In the case of a classification tree, there are three methods of splitting: Gini index, entropy loss, and classification error. Since they deal with classification problems, we will study these criteria in the next chapter.

For a regression tree, variance reduction is the criteria for splitting. In variance splitting, variance at each node is calculated using the following formula.

$$\text{Variance} = \frac{\Sigma(x - \overline{x})^2}{n} \qquad \text{(Equation 2-15)}$$

We calculate variance for each split as the weighted average of variance for each node. And then the split with a lower variance is selected for the purpose of splitting.

There are quite a few decision tree algorithms available, like ID3, CART, C4.5, CHAID, and so on. These algorithms are explored in more detail in Chapter 3 after we have discussed concepts of classification using a decision tree.

Case study: Petrol consumption using Decision tree

It is time to develop a Python solution using a decision tree. The code and dataset are uploaded to the Github repository. You are advised to download the dataset from the Github link shared at the start of the chapter.

Step 1: Import all the libraries first.

```
import pandas as pd
import numpy as np
import matplotlib.pyplot as plt
%matplotlib inline
```

Step 2: Import the dataset using read_csv file command.

```
petrol_data = pd.read_csv('petrol_consumption.csv')
petrol_data.head(5)
```

	Petrol_tax	Average_income	Paved_Highways	Population_Driver_licence(%)	Petrol_Consumption
0	9.0	3571	1976	0.525	541
1	9.0	4092	1250	0.572	524
2	9.0	3865	1586	0.580	561
3	7.5	4870	2351	0.529	414
4	8.0	4399	431	0.544	410

Step 3: Let's explore the major metrices of the independent variables:

```
petrol_data.describe()
```

	Petrol_tax	Average_income	Paved_Highways	Population_Driver_licence(%)	Petrol_Consumption
count	48.000000	48.000000	48.000000	48.000000	48.000000
mean	7.668333	4241.833333	5565.416667	0.570333	576.770833
std	0.950770	573.623768	3491.507166	0.055470	111.885816
min	5.000000	3063.000000	431.000000	0.451000	344.000000
25%	7.000000	3739.000000	3110.250000	0.529750	509.500000
50%	7.500000	4298.000000	4735.500000	0.564500	568.500000
75%	8.125000	4578.750000	7156.000000	0.595250	632.750000
max	10.000000	5342.000000	17782.000000	0.724000	968.000000

Step 4: We will now split the data into train and test and then try to fit the model. First the X and y variables are segregated and then they are split into train and test. Test is 20% of training data.

```
X = petrol_data.drop('Petrol_Consumption', axis=1)
y = petrol_data ['Petrol_Consumption']
from sklearn.model_selection import train_test_split
X_train, X_test, y_train, y_test = train_test_split(X, y,
test_size=0.2, random_state=0)
from sklearn.tree import DecisionTreeRegressor
```

```
decision_regressor = DecisionTreeRegressor()
decision_regressor.fit(X_train, y_train)
```

Step 5: Use the model for making a prediction on the test dataset.

```
y_pred = decision_regressor.predict(X_test)
df=pd.DataFrame({'Actual':y_test, 'Predicted':y_pred})
df
```

	Actual	Predicted
29	534	547.0
4	410	414.0
26	577	574.0
30	571	554.0
32	577	631.0
37	704	644.0
34	487	648.0
40	587	649.0
7	467	414.0
10	580	498.0

Step 6: Measure the performance of the model created by using various measuring parameters.

```
from sklearn import metrics
print('Mean Absolute Error:', metrics.mean_absolute_error
(y_test, y_pred))
print('Mean Squared Error:', metrics.mean_squared_error
(y_test, y_pred))
print('Root Mean Squared Error:', np.sqrt(metrics.mean_squared_
error(y_test, y_pred)))
```

```
Mean Absolute Error: 50.9
Mean Squared Error: 4629.7
Root Mean Squared Error: 68.0418988565134
```

The MAE for our algorithm is 50.9, which is less than 10% of the mean of all the values in the 'Petrol_Consumption' column, indicating that our algorithm is doing a good job.

The visualization of the preceding solution is the code is checked in at GitHub link.

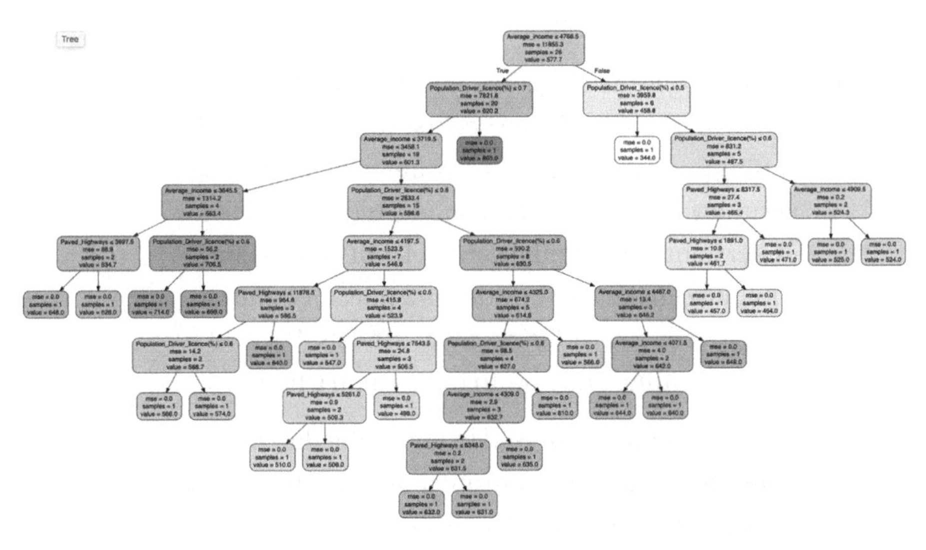

This is the Python implementation of a decision tree. The code can be replicated for any problem we want to solve using a decision tree. We will now explore the pros and cons of decision tree algorithms.

Advantages of Decision Tree

1. Decision trees are easy to build and comprehend. Since they mimic human logic in decision making, the output looks very structured and easy to grasp.

2. They require very less data preparation. They are able to work for both regression and classification problems and can handle huge datasets.

3. Decision trees are not impacted much by collinearity of the variables. The significant variable identification is inbuilt and we validate the outputs of decision trees using statistical tests.

4. Perhaps the most important advantage of decision trees is that they are very intuitive. Stakeholders or decision makers who are not from a data science background can also understand the tree.

Disadvantages of Decision Tree

1. Overfitting is the biggest problem faced in the decision tree. Overfitting occurs when the model is getting good training accuracy but very low testing accuracy. It means that the model has been able to understand the training data well but is struggling with unseen data. Overfitting is a nuisance and we have to reduce overfitting in our models. We deal with overfitting and how to reduce it in Chapter 5.

2. A greedy approach is used to create the decision trees. Hence, it may not result in the best tree or the globally optimum tree. There are methods proposed to reduce the impact of the greedy algorithm like *dual information distance* (DID). The DID heuristic makes a decision on attribute selection by considering both immediate and future potential effects on the overall solution. The classification tree is constructed by searching for the shortest paths over a graph of partitions. The shortest path identified is defined by the selected features. The DID method considers

a. The orthogonality between the selected partitions,

b. The reduction of uncertainty on the class partition given the selected attributes.

3. They are quite "touchy" to the training data changes and hence sometimes a small change in the training data can lead to a change in the final predictions.

4. For classification trees, the splitting is biased towards the variable with the higher number of classes.

We have discussed the concepts of decision tree and developed a case study using Python. It is very easy to comprehend, visualize, and explain. Everyone is able to relate to a decision tree, as it works in the way we make our decisions. We choose the best parameter and direction, and then make a decision on the next step. Quite an intuitive approach!

This brings us towards the end of decision tree–based solutions. So far we have discussed simple linear regression, multinomial regression, nonlinear regression, and decision tree. We understood the concepts, the pros and cons, and the assumptions, and we developed respective solutions using Python. It is a very vital and relevant step towards ML.

But all of these algorithms work individually and one at a time. It allows us to bring forward the next generation of solutions called *ensemble methods*, which we will examine now.

Ensemble Methods for Regression

"United we stand" is the motto for ensemble methods. They use multiple predictors and then "unite" or collate the information to make a final decision.

Formally put, ensemble methods train multiple predictors on a dataset. These predictor models might or might not be weak predictors themselves individually. They are selected and trained in such a way that each has a slightly different training dataset and may get slightly different results. These individual predictors might learn a different pattern from each other. Then finally, their individual predictions are combined and a final decision is made. Sometimes, this combined group of learners is referred to as *meta* model.

In ensemble methods, we ensure that each predictor is getting a slightly different data set for training. This is usually achieved at random with replacement or *bootstrapping*. In a different method, we can adjust the weights assigned to each of the data points. This increases the weights, that is, the focus on those data points.

To visualize how ensemble methods make a prediction, we can consider Figure 2-22 where a random forest is shown and predictor models are trained on slightly different datasets.

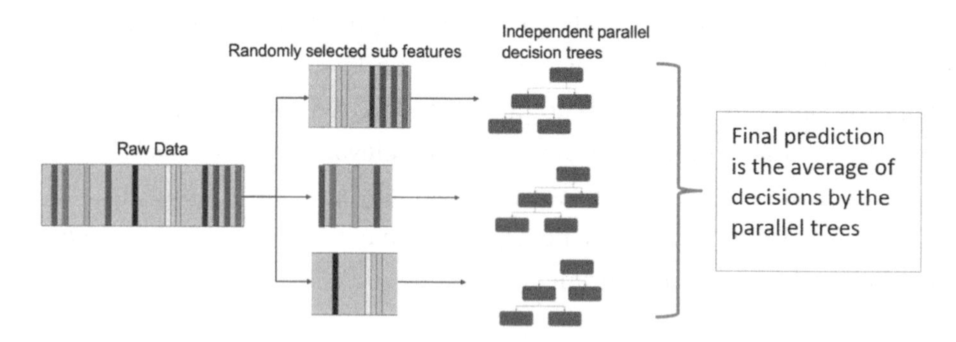

Figure 2-22. *Ensemble learning-based random forest where raw data is split into randomly selected subfeatures and then individual independent parallel trees are created. The final result is the average of all the predictions by subtrees.*

Ensemble methods can be divided into two broad categories: bagging and boosting.

1. **Bagging models** or bootstrap aggregation improves the overall accuracy by the means of several weak models. The following are major attributes for a bagging model:

    a. Bagging uses sampling with replacement to generate multiple datasets.

    b. It builds multiple predictors simultaneously and independently of each other.

    c. To achieve the final decision an average/vote is done. It means if we are trying to build a regression model, the average or median of all the respective predictions will be taken while for the classification model a voting is done.

    d. Bagging is an effective solution to tackle variance and reduce overfitting.

    e. Random forest is one of the examples of a bagging method (as shown in Figure 2-22).

2. **Boosting**: Similar to bagging, boosting also is an ensemble method. The following are the main points about boosting algorithm:

    a. In boosting, the learners are grown sequentially from the last one.

    b. Each subsequent learner improves from the last iteration and focuses more on the errors in the last iteration.

c. During the process of voting, higher vote is awarded to learners which have performed better.

d. Boosting is generally slower than bagging but mostly performs better.

e. Gradient boosting, extreme gradient boosting, and AdaBoosting are a few example solutions.

It is time for us to develop a solution using random forest. We will be exploring more on boosting in Chapter 4, where we study supervised classification algorithms.

Case study: Petrol consumption using Random Forest

For a random forest regression problem, we will be using the same case study we used for decision tree. In the interest of space, we are progressing after creating the training and testing dataset

Step 1: Import all the libraries and the dataset. We have already covered the steps while we were implementing decision tree algorithms.

```
import pandas as pd
import numpy as np
import matplotlib.pyplot as plt
%matplotlib inline
petrol_data = pd.read_csv('petrol_consumption.csv')
X = petrol_data.drop('Petrol_Consumption', axis=1)
y = petrol_data['Petrol_Consumption']
from sklearn.model_selection import train_test_split
X_train, X_test, y_train, y_test = train_test_split(X, y,
test_size=0.20, random_state=0)
```

Step 2: Import the random forest regressor library and initiate a RandomForestRegressor variable.

```
from sklearn.ensemble import RandomForestRegressor
randomForestModel = RandomForestRegressor(n_estimators=200,
                                    bootstrap = True,
                                    max_features = 'sqrt')
```

Step 3: Now fit the model on training and testing data.

```
randomForestModel.fit(X_train, y_train)
```

Step 4: We will now predict the actual values and check the accuracy of the model.

```
rf_predictions = randomForestModel.predict(X_test)
from sklearn import metrics
print('Mean Absolute Error:', metrics.mean_absolute_error
(y_test, rf_predictions))
print('Mean Squared Error:', metrics.mean_squared_error
(y_test, rf_predictions))
print('Root Mean Squared Error:', np.sqrt(metrics.mean_squared_
error(y_test, rf_predictions)))
```

```
Mean Absolute Error: 58.5
Mean Squared Error: 5273.9
Root Mean Squared Error: 72.62162212454359
```

Step 5: Now we will extract the two most important features. Get the list of all the columns present in the dataset and we will get the numeric feature importance.

```
feature_list=X_train.columns
importances = list(randomForestModel.feature_importances_)
```

```
feature_importances = [(feature, round(importance, 2)) for
feature, importance in zip(feature_list, importances)]
feature_importances = sorted(feature_importances, key = lambda
x: x[1], reverse = True)
[print('Variable: {:20} Importance: {}'.format(*pair)) for pair
in feature_importances];
```

```
Variable: Average_income        Importance: 0.29
Variable: Paved_Highways        Importance: 0.28
Variable: Population_Driver_licence(%) Importance: 0.28
Variable: Petrol_tax            Importance: 0.15
```

Step 6: We will now re-create the model with important variables.

```
rf_most_important = RandomForestRegressor(n_estimators= 500,
random_state=5)
important_indices = [feature_list[2], feature_list[1]]
train_important = X_train.loc[:, ['Paved_Highways','Average_
income','Population_Driver_licence(%)']]
test_important = X_test.loc[:, ['Paved_Highways','Average_
income','Population_Driver_licence(%)']]
train_important = X_train.loc[:, ['Paved_Highways','Average_
income','Population_Driver_licence(%)']]
test_important = X_test.loc[:, ['Paved_Highways','Average_
income','Population_Driver_licence(%)']]
```

Step 7: Train the random forest algorithm.

```
rf_most_important.fit(train_important, y_train)
```

Step 8: Make predictions and determine the error.

```
predictions = rf_most_important.predict(test_important)
predictions
```

Step 9: Print the mean absolute error, mean squared error, and root mean squared error.

```
print('Mean Absolute Error:', metrics.mean_absolute_error
(y_test, predictions))
print('Mean Squared Error:', metrics.mean_squared_error(y_test,
predictions))
print('Root Mean Squared Error:', np.sqrt(metrics.mean_squared_
error(y_test, predictions)))
```

```
Mean Absolute Error: 54.9088
Mean Squared Error: 4297.8499098
Root Mean Squared Error: 65.55798890905669
```

As we can observe, after selecting the significant variables the error has reduced for random forest.

Ensemble learning allows us to collate the power of multiple models and then make a prediction. These models individually are weak but together act as a strong model for prediction. And that is the beauty of ensemble learning. We will now discuss pros and cons of ensemble learning.

Advantages of ensemble learning:

1. An ensemble model can result in lower variance and low bias. They generally have a better understanding of the data.
2. The accuracy of ensemble methods is generally higher than regular methods.
3. Random forest model is used to tackle overfitting, which is generally a concern for decision trees. Boosting is used for bias reduction.
4. And most importantly, ensemble methods are a collection of individual models. Hence, more complex understanding of the data is generated.

Challenges with ensemble learning:

1. Owing to the complexity of ensemble learning, it is difficult to comprehend. For example, while we can easily visualize a decision tree it is difficult to visualize a random forest model.

2. Complexity of the models does not make them easy to train, test, deploy, and refresh, which is generally not the case with other models.

3. Sometimes, ensemble models take a long time to converge and train. And that increases the training time.

We have covered the concept of ensemble learning and developed a regression solution using random forest. Ensemble learning has been popular for a long time and has won quite a few competitions in Kaggle. You are advised to understand the concepts and replicate the code implementation.

Before we close the discussion on ensemble learning, we have to study an additional concept of feature selection using decision tree. Recall from the last section where we developed a multiple regression problem; we will be continuing with the same problem to select the significant variables.

Feature Selection Using Tree-Based Methods

We are continuing using the dataset we used in the last section where we developed a multiple regression solution using house data. We are using ensemble-based ExtraTreeClassifier to select the most significant features.

The initial steps of importing the libraries and dataset remain the same.

Step 1: Import the libraries and dataset.

```
import numpy as np
import pandas as pd
import matplotlib.pyplot as plt
%matplotlib inline
feature_df = pd.read_csv('House_data.csv')
```

Step 2: Perform similar data preprocessing we did in the regression problem.

```
feature_df['basement'] = (feature_df['sqft_basement'] > 0).
astype(int)
feature_df['renovated'] = (feature_df['yr_renovated'] > 0).
astype(int)
to_drop = ['id', 'date', 'sqft_basement', 'yr_renovated']
feature_df.drop(to_drop, axis = 1, inplace = True)
cat_cols = ['waterfront', 'view', 'condition', 'grade',
'floors']
feature_df = pd.get_dummies(feature_df, columns = cat_cols,
drop_first=True)
y = feature_df.iloc[:, 0].values
X = feature_df.iloc[:, 1:].values
```

Step 3: Let us now create a ExtraTreeClassifier.

```
from sklearn.ensemble import ExtraTreesClassifier
tree_clf = ExtraTreesClassifier()
tree_clf.fit(X, y)
tree_clf.feature_importances_
```

Step 4: We are now getting the respective importance of various variables and ordering them in descending order of importance.

```
importances = tree_clf.feature_importances_
feature_names = feature_df.iloc[:, 1:].columns.tolist()
feature_names
feature_imp_dir = dict(zip(feature_names, importances))
features = sorted(feature_imp_dir.items(), key=lambda x: x[1],
reverse=True)
feature_imp_dir
```

Step 5: We will visualize the features in the order of their importance.

```
plt.bar(range(len(features)), [imp[1] for imp in features],
align='center')
plt.title('The important features in House Data');
```

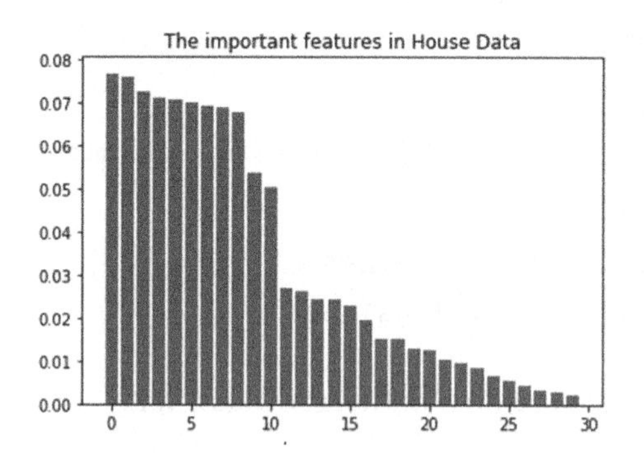

Step 6: We will now analyze how many variables have been selected and how many have been removed.

```
from sklearn.feature_selection import SelectFromModel
abc = SelectFromModel(tree_clf, prefit = True)
x_updated = abc.transform(X)
print('Total Features count:', np.array(X).shape[1])
print('Selected Features: ',np.array(x_updated).shape[1])
```

The output shows that the total number of variables was 30; from that list, 11 variables have been found to be significant.

Using ensemble learning–based ExtraTreeClassifier is one of the techniques to shortlist significant variables. We can look at the respective p-values and shortlist the variables.

Ensemble learning is a very powerful method to combine the power of weak predictors and make them strong enough to make better predictions. It offers a fast, easy, and flexible solution which is applicable for both classification and regression problems. Owing to their flexibility sometimes we encounter the problem of overfitting, but bagging solutions like random forest tend to overcome the problem of overfitting. Since ensemble techniques promote diversity in the modeling approach and use a variety of predictors to make the final decision, many times they have outperformed classical algorithms and hence have gained a lot of popularity.

This brings us to the end of ensemble learning methods. We are going to revisit them in Chapter 3 and Chapter 4.

So far, we have covered simple linear regression, multiple linear regression, nonlinear regression, decision tree, and random forest and have developed Python codes for them too. It is time for us to move to the summary of the chapter.

Summary

We are harnessing the power of data in more innovative ways. Be it through reports and dashboards, visualizations, data analysis, or statistical modeling, data is powering the decisions of our businesses and processes. Supervised regression learning is quickly and quietly impacting the decision-making process. We are able to predict the various indicators of our business and take proactive measures for them. The use cases are across pricing, marketing, operations, quality, Customer Relationship Management (CRM), and in fact almost all business functions.

And regression solutions are a family of such powerful solutions. Regression solutions are very profound, divergent, robust, and convenient. Though there are fewer use cases of regression problems as compared to classification problems, they still serve as the foundation of supervised learning models. Regression solutions are quite sensitive to outliers and changes in the values of target variables. The major reason is that the target variable is continuous in nature.

Regression solutions help in modeling the trends and patterns, deciphering the anomalies, and predicting for the unseen future. Business decisions can be more insightful in light of regression solutions. At the same time, we should be cautious and aware that the regression will not cater to unseen events and values which are not trained for. Events such as war, natural calamities, government policy changes, macro/micro economic factors, and so on which are not planned will not be captured in the model. We should be cognizant of the fact that any ML model is dependent on the quality of the data. And for having an access to clean and robust dataset, an effective data-capturing process and mature data engineering is a prerequisite. Then only can the real power of data be harnessed.

In the first chapter, we learned about ML, data and attributes of data quality, and various ML processes. In this second chapter, we have studied regression models in detail. We examined how a model is created, how we can assess the model's accuracy, pros and cons of the model, and implementation in Python too. In the next chapter, we are going to work on supervised learning classification algorithms.

You should be able to answer the following questions.

EXERCISE QUESTIONS

Question 1: What are regression and use cases of regression problem?

Question 2: What are the assumptions of linear regression?

Question 3: What are the pros and cons of linear regression?

Question 4: How does a decision tree make a split for a node?

Question 5: How does an ensemble method make a prediction?

Question 6: What is the difference between the bagging and boosting approaches?

Question 7: Load the data set Iris using the following command:

```
from sklearn.datasets import load_iris
iris = load_iris()
```

Predict the sepal length of the iris flowers using linear regression and decision tree and compare the results.

Question 8: Load the auto-mpg.csv from the Github link and predict the mileage of a car using decision tree and random forest and compare the results. Get the most significant variables and re-create the solution to compare the performance.

Question 9: The next dataset contains information about the used cars listed on `www.cardekho.com`. It can be used for price prediction. Download the dataset from `https://www.kaggle.com/nehalbirla/vehicle-dataset-from-cardekho`, perform the EDA, and fit a linear regression model.

Question 10: The next dataset is a record of seven common species of fish. Download the data from `https://www.kaggle.com/aungpyaeap/fish-market` and estimate the weight of fish using regression techniques.

Question 11: Go through the research papers on decision trees at https://ieeexplore.ieee.org/document/9071392 and https://ieeexplore.ieee.org/document/8969926.

Question 12: Go through the research papers on regression at https://ieeexplore.ieee.org/document/9017166 and https://agupubs.onlinelibrary.wiley.com/doi/full/10.1002/2013WR014203.

CHAPTER 3

Supervised Learning for Classification Problems

"Prediction is very difficult, especially if it's about the future."

— Niels Bohr

We live in a world where the predictions for an event help us modify our plans. If we know that it is going to rain today, we will not go camping. If we know that the share market will crash, we will hold on our investments for some time. If we know that a customer will churn out from our business, we will take some steps to lure them back. All such predictions are really insightful and hold strategic importance for our business.

It will also help if we know the factors which make our sales go up/down or make an email work or result in a product failure. We can work on our shortcomings and continue the positives. The entire customer-targeting strategy can be modified using such knowledge. We can revamp the online interactions or change the product testing: the implementations are many. Such ML models are referred to as *classification* algorithms, the focus of this chapter.

V. Verdhan, *Supervised Learning with Python*,
https://doi.org/10.1007/978-1-4842-6156-9_3

In Chapter 2, we studied the regression problems which are used to predict a continuous variable. In this third chapter we will examine the concepts to predict a categorical variable. We will study how confident we are for an event to happen or not. We will study logistic regression, decision tree, k-nearest neighbor, naïve Bayes, and random forest in this chapter. All the algorithms will be studied; we will also develop Python code using the actual dataset. We will deal with missing values, duplicates, and outliers, do an EDA, measure the accuracy of the algorithm, and choose the best algorithm. Finally, we will solve a case study to complete the understanding.

Technical Toolkit Required

We are using Python 3.5+ in this book. You are advised to get Python installed on your machine. We are using Jupyter notebook; installing Anaconda-Navigator is required for executing the codes.

The major libraries used are numpy, pandas, matplotlib, seaborn, scikitlearn, and so on. You are advised to install these libraries in your Python environment. All the codes and datasets have been uploaded at the Github repository at the following link: `https://github.com/Apress/supervised-learning-w-python/tree/master/Chapter%203`

Before starting with classification "machine learning," it is imperative to examine the statistical concept of critical region and p-value, which we are studying now. They are useful to judge the significance for a variable out of all the independent variables. A very strong and critical concept indeed!

Hypothesis Testing and p-Value

Imagine a new drug X is launched in the market, which claims to cure diabetes in 90% of patients in 5 weeks. The company tested on 100 patients and 90 of them got cured within 5 weeks. How can we ensure that the drug is indeed effective or if the company is making false claims or the sampling technique is biased?

Hypothesis testing is precisely helpful in answering the questions asked.

In hypothesis testing, we decide on a hypothesis first. In the preceding example of a new drug, our hypothesis is that drug X cures diabetes in 90% of patients in 5 weeks. That is called the *null hypothesis* and is represented by H_0. In this case, H_0 is 0.9. If the null hypothesis is rejected based on evidence, an *alternate hypothesis* H_1 needs to be accepted. In this case, $H_1 < 0.9$. We always start with assuming that the null hypothesis is true.

Then we define our significance level, α. It is a measure for how unlikely we want the results of the sample to be, before we reject the null hypothesis H_0. Refer to Figure 3-1(i) and you will be able to relate to the understanding.

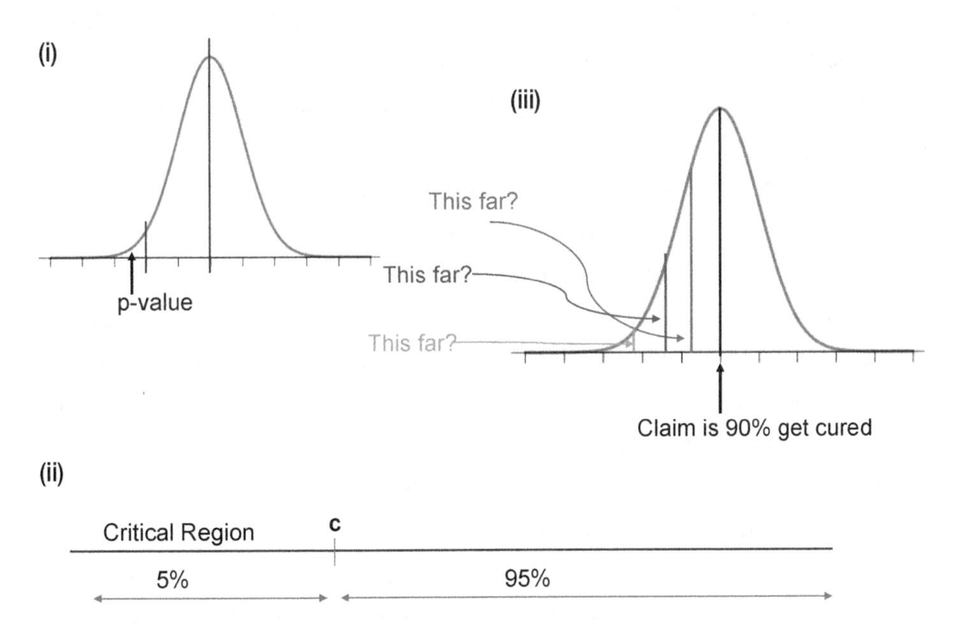

Figure 3-1. *(i) The significance level has to be specified, that is, until what point we will accept the results and not reject the null hypothesis. (ii) The critical region is the 5% region shown in the middle image. (iii) The right side image is showing the p-value and how it falls in the critical region in this case.*

We then define the critical region as "c" as shown in Figure 3-1(ii).

If X represents the number of diabetic patients cured, the critical region is defined as $P(X<c) < \alpha$ where $\alpha = 5\%$. In a 95% confidence interval, there is a 5% chance that the sample will not have the population mean. It is interpreted as, if a sample we are interested in falls in a critical region then we can safely reject the null hypothesis.

This is precisely the reason for 5% or 0.05 being referred to as the *significance level.* If we want a confidence interval of 99%, then 0.01 is the significance level.

Then the next step is to get the p-value. Refer to Figure 3-1(iii) for better understanding.

Formally out, *p-value* is the probability of getting a value up to and including the one in the sample in the direction of the critical region. It is a method to check if the results of a sample fall in the critical region of the hypothesis test. So, based on the p-value we take a call to reject or accept the null hypothesis.

Once the p-value is calculated, we analyze if the value falls in the critical region. If the answer we get is yes, we can reject the null hypothesis.

This knowledge of hypothesis testing and p-value is critical as it paves the way for identifying the *significant variables.* Whenever we train our ML algorithms, along with other results we get a p-value for each of the variables. As a rule of thumb, if the p-value is less than or equal to 0.05, the variable is considered as significant.

In this case, the null hypothesis is that the independent variable is not significant and does not impact the target variable. If the p-value is less than or equal to 0.05, we can reject the null hypothesis and hence can conclude that the variable is indeed significant. In the language of statistics, if the p-value for an independent variable x is less than or equal to 0.05, it suggests strong evidence against the null hypothesis as there is less than a 5% chance of the null hypothesis being correct. Or in other words, the variable x is a significant variable to make a prediction for the target variable y. But it does not mean there is 95% probability for the alternate hypothesis to be correct. To be noted is that p-value is a condition upon the null hypothesis and is unrelated to the alternate hypothesis.

We use the p-value to shortlist the significant variables and compare their respective importance. It is a universally accepted metric to choose significant variables.

We will now proceed to the concepts of classification algorithms in the next section.

Classification Algorithms

In our day-to-day business, we make decisions to either invest in stock or not, send a communication to a customer or not, accept a product or reject it, accept an application or ignore it. The basis of these decisions is some sort of insight we have about our business, our processes, our goals, and the factors which enter into our decision making. At the same time, we do expect a favorable output from this decision of ours.

Supervised classified algorithms are used to generate such insights and help us take that decision. They predict the probability for an event to happen or not. At the same time, depending on the choice of supervised learning algorithm, we get to know the factors which impact the occurrences of such an event.

Formally put, *classification* algorithms are a branch of supervised ML algorithms which are used to model the probability for a certain class. For example, if we want to perform binary classification, we will be modeling for two classes like pass or fail, healthy or sick, fraudulent or genuine, yes or no, and so on. It can be extended to multiclass classification problems—for example, good/bad/neutral, red/yellow/blue/green, cat/dog/horse/car, and so on.

Like a regression model, there is a target variable and independent variables. The only difference is that the target variable is categorical in nature. The independent variables used to make the predictions can be continuous or categorical, and it depends on the algorithm used for modeling.

The following use cases will make the usage of classification algorithms clear:

1. A retailer is losing its repeat customers, that is, customers who used to make purchases are not coming back. The supervised learning algorithm will help identify the customers who are more prone to churn and not come back. The retailer can then target those customers selectively and can offer discounts to bring them back to the business.

2. A manufacturing plant has to maintain the best quality of their products. And for that the technical team would like to ascertain if a particular combination of tools, raw materials, and physical conditions will lead to the best quality and yield. Supervised algorithms can help in that selection.

3. An insurance provider wants to model which customers should be given the policy or not. Depending on the customers' previous history, employment details, transaction patterns, and so on, the decision has to be made. Here the classification ML model can help to predict acceptance score for the customers, which can be used to accept or reject the application.

4. A bank offering credit cards to its customers has to identify which incoming transactions are fraudulent and which are genuine. Based on the transaction details like origin, amount, time of transaction, mode, and other customer parameters, a decision has to be made. Supervised classification algorithms will be helpful to making that decision.

5. A telecom operator wishes to launch a new data product in the market. For it, the need is to target a few subscribers who have a higher probability of being interested in the product and recharging with it. Supervised classification algorithms will be able to generate a score for each subscriber and subsequently an offer can be made.

The preceding use cases are some of the pragmatic implementations in the industry. A classification algorithm generates a probability score for an event and accordingly the sales/marketing/operations/quality/risk teams can take a business call. Quite a powerful usage and very handy too!

There are quite a few algorithms which serve the purpose and we will be discussing a few in this chapter and rest in Chapter 4.

The algorithms which can be used are

- Logistic regression
- k-nearest neighbor
- Decision tree
- Random forest
- Naïve Bayes
- SVM

- Gradient boosting
- Neural networks

We are discussing the first five algorithms in this chapter and rest in the next chapter. Let us start the discussion with the logistic regression algorithm in the next section.

Logistic Regression for Classification

In Chapter 2 we learned how to predict the value for a continuous variable like number of customers, sales, rainfall, and so on using linear regression. Now we have to predict whether a customer will visit or not, whether the sale will go up or not, and so on. Using logistic regression, we can model and solve the preceding problems.

Formally put, logistic regression is a statistical model that utilizes logit function to model classification problems, that is, a categorical dependent variable. In the basic form, we model a *binary* classification problem and refer to it as binary logistic regression. In complex problems, where more than one categories have to be modeled, we use *multinomial* logistic regression.

Let us understand logistic regression by means of an example.

Consider we have to make a decision whether a credit card transaction is fraudulent or not. The response is binary (Yes or No); if the transaction seems promising it will be accepted, otherwise not. We will have incoming transaction attributes like amount, time of transaction, payment mode, and so on, and we have to make a decision based on them.

For such a problem, logistic regression models the probability of fraud. In the preceding case, the probability of fraud can be

$$\text{Probability (fraud} = \text{Yes} \mid \text{amount)}$$

The value for this probability will lie between 0 and 1. It can be interpreted as follows: given a value of "fraud amount" we can make a prediction for the genuineness of a credit card transaction.

The question arises of how we model such a relationship. If we use a *linear* equation, we can simply write it as

$$\text{Probability or } pr(x) = \beta_0 + \beta_1 x \qquad \text{(Equation 3-1)}$$

If we want to predict for "success" using "amount" by fitting the preceding formula, it can be represented in the form of the following graph. For the smaller values of the fraud amount, we can see that probability is less than zero while for the large values, fraud probability is greater than one. And both of these situations are not possible.

Hence, logistic regression is used to tackle this problem. Logistic regression uses the *sigmoid* function, which takes input as any real value and gives an output between 0 and 1. The standard *logistic regression* equation can be represented as in Equation 3-2 and shown in Figure 3-2.

$$P(x) = \frac{e^t}{(e^t + 1)} \text{ where } t = \beta_0 + \beta_1 x \qquad \text{(Equation 3-2)}$$

which can be rewritten as

$$P(x) = \frac{e^{\beta_0 + \beta_1 x}}{(1 + e^{\beta_0 + \beta_1 x})} \qquad \text{(Equation 3-3)}$$

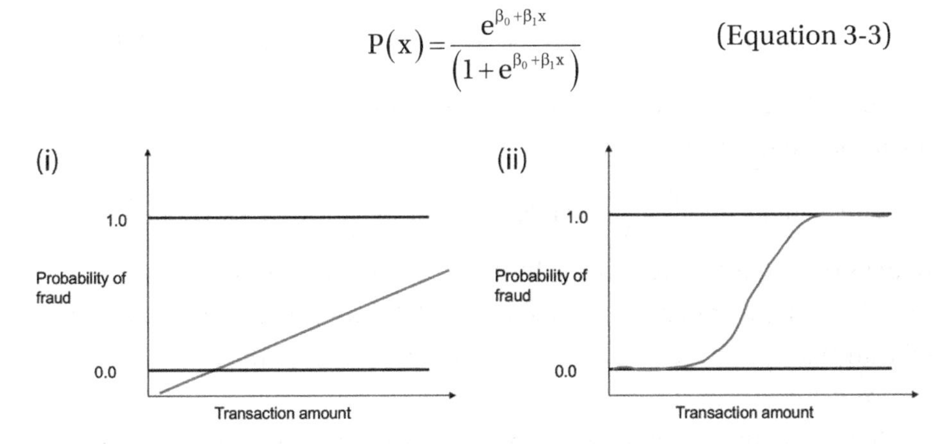

Figure 3-2. *(i) Linear regression function will not be able to do justice to the task of predicting fraud. (ii) Logistic regression having an "S"-shaped curve will be more suitable as it will give scores between 0 and 1.*

The next question that comes to mind is how to fit this equation. Recall in the case of linear regression, we want to make a prediction of the target variable as close to the actual values. A similar approach is followed here too. The fitting of the logistic regression is done using the *maximum likelihood* function. The likelihood function measures the goodness of fit of a statistical model and for logistic regression sometimes referred to as log-likelihood function. The mathematical proof is beyond the scope of the book.

If we manipulate Equation 3-3 we will get

$$\log\left(\frac{p(x)}{1-p(x)}\right)=\beta_0+\beta_1 x \qquad \text{(Equation 3-4)}$$

If we take a natural log of both the sides,

$$\left(\frac{p(x)}{1-p(x)}\right)=e^{\beta_0+\beta_1 x} \qquad \text{(Equation 3-5)}$$

In Equation 3-5, the quantity $\frac{p(x)}{(1-p(x))}$ is referred to as *odds*. It is called odds because it is more intuitive as compared to probability, as odds are a common jargon in betting.

The term $\log\left(\frac{p(x)}{1-p(x)}\right)$ is called *logit*. If we compare with a linear regression equation, we can easily make out that with each unit increase in x, the logit (or sometimes called log-odds) changes by β_1. This value can take any value between 0 and infinity. We can visualize the function in Figure 3-2(ii).

In most of the business problems, we have more than one independent variable and hence we use multinomial logistic regression to solve it. Mathematically, it can be represented as follows:

$$\log\left(\frac{p(x)}{1-p(x)}\right)=\beta_0+\beta_1 x+\ldots+\beta_n \qquad \text{(Equation 3-6)}$$

But there is a question which still remains unanswered: why do we need logistic regression when we have linear regression with us?

Let's say a bank is making an assessment of its service quality, based on the customer's historical transactions and service details. In such a case, the predicted response is going to be positive, negative, or neutral. We code these responses as

Target variable y = 1 for positive,
2 for negative,
3 for neutral

} We have three categories of responses over here

If we treat the target variable as a continuous variable, it means we have to predict the actual value of y. But this implies that positive is one less than negative and negative is one less than neutral. And the difference between positive and negative is the same as the difference between negative and neutral. This argument is intrinsically wrong and does not make any practical sense.

Moreover, even if we reduce the number of responses from three to two, let's say positive and negative only, then too the linear regression might give us some probability score beyond 1 or less than 0, which is mathematically not possible. If we fit the best-found regression line, it still won't be enough to decide any point by which we can differentiate between the two classes. It will classify some positive as negative and vice versa. Moreover, if we get a score of 0.5 from linear regression, should it be classified as positive or negative? And an outlier can completely mess the outputs for us. Hence it is practically more sensible to use a classification algorithm like logistic regression instead of a linear regression model to solve the problem.

Like linear regression, logistic regression has a few assumptions:

1) Being a classification algorithm, the outcome of the logistic regression model is a binary or dichotomous variable like success/fail, yes/no, or zero/one.

2) There exists a linear relationship between the logit of the outcome and each of the independent variables.

3) Outliers do not exist or at least there are no significant outliers for continuous variables.

4) There exists no correlation between the independent variables.

An important point to be noted is that the accuracy of the algorithm depends on the training data which has been used to train the algorithm. If the training data is not representative, then the resultant model will not be a robust one. The training data should conform to the data quality standards we discussed in Chapter 1. We will be revisiting this concept in detail in Chapter 5.

Tip To be able to have a representative dataset, it is advisable to have a minimum of 10 data points for each of the independent variables with reference to their least frequent value. For example, for 20 independent variables and a least frequent outcome of 0.2, we should have (20*10)/0.2 = 1000 data points.

Key points to note about logistic regression:

1) The output of a logistic classification model generally is a probability score for an event. It can be used for both binary classification and multi classification problems.

2) Since the output is probability, it cannot go beyond 1 and cannot be less than 1. And hence the shape of the logistic curve is "S".

3) It can handle any number of classes as the target variable as well as both categorical and continuous independent variables.

4) The maximum likelihood algorithm helps to determine the respective coefficients for the equation. It is not required for the independent variables to be normally distributed or have an equal variance in each group.

5) $\frac{P}{1\text{-}p}$ is the odds ratio and whenever this value is positive, the chances of success are above 50%.

6) The interpretation of coefficients is difficult in logistic regression as the relation is not straightforward as in the case of linear regression.

Before moving further, it is imperative to carefully examine the accuracy measurement methods. You are advised to be thorough with each of them. A vital component in supervised learning, indeed!

Assessing the Accuracy of the Solution

The objective to create an ML solution is to predict for future events. But before deploying the model to a production environment, it is imperative we measure the performance of the model. Moreover, we generally train multiple algorithms with multiple iterations. We have to choose the best algorithm based on the variously accurate KPIs. In this section, we are studying the most important accuracy assessment criteria.

The most important measures to measure the efficacy of a classification problem are as follows:

1. **Confusion matrix**: One of the most popular methods is *confusion matrix*. It can be used for both binary and multiclass problems. In its simplest form it is represented as a 2×2 matrix in Figure 3-3.

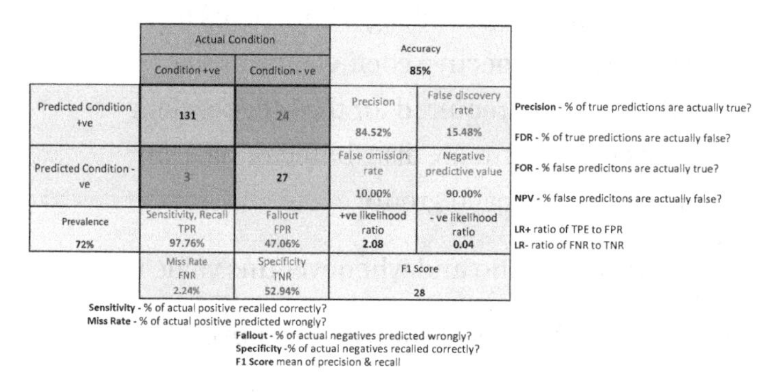

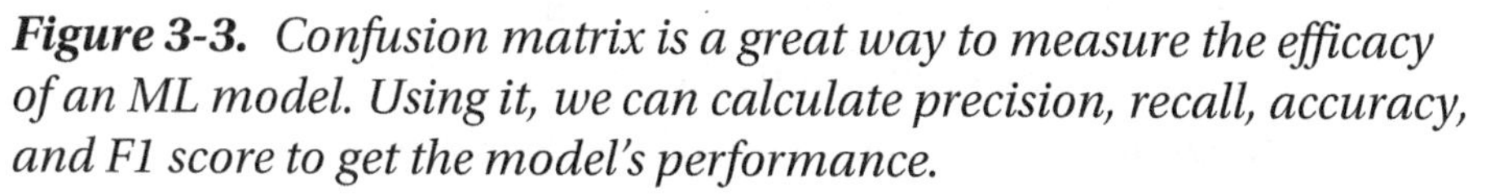

***Figure 3-3.** Confusion matrix is a great way to measure the efficacy of an ML model. Using it, we can calculate precision, recall, accuracy, and F1 score to get the model's performance.*

We will now learn about each of the parameters separately:

a. **Accuracy**: Accuracy is how many predictions were made correctly. In the preceding example, the accuracy is (131+27)/ (131+27+3+24) = 85%

b. **Precision**: Precision represents out of positive predictions; how many were actually positive. In the preceding example, precision is 131/ (131+24) = 84%

c. **Recall or sensitivity**: Recall is how of all the actual positive events; how many we were able to capture. In this example, 131/ (131+3) = 97%

d. **Specificity or true negative rate**: Specificity is out of actual negatives; how many were predicted correctly. In this example: 27/ (27+24) = 92%

2. **ROC curve and AUC value**: ROC or receiver operating characteristics is used to compare different models. It is a plot between TPR (true positive rate) and FPR (false positive rate). The area under the ROC curve (AUC) is a measure of how good a model is. The higher the AUC values, the better the model, as depicted in Figure 3-4. The straight line at the angle of 45° represents 50% accuracy. A good model is having an area above 0.5 and hugging to the top left corner of the graph as shown in Figure 3-4. The one in the green seems to be the best model here.

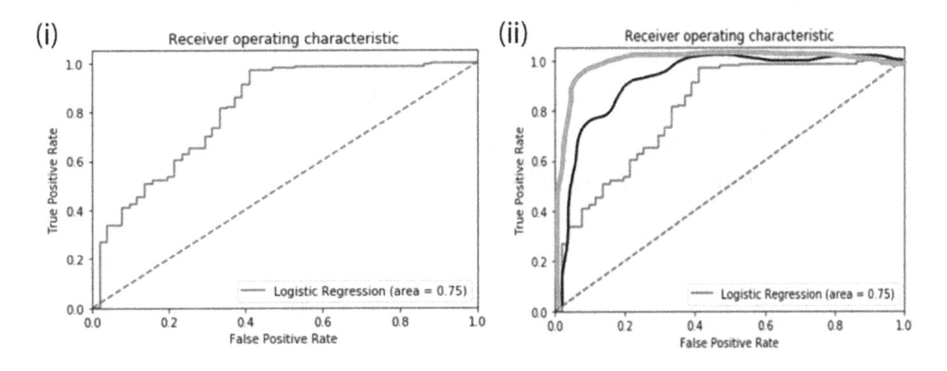

Figure 3-4. *(i) ROC curve is shown on the left side. (ii) Different ROC curves are shown on the right. The green ROC curve is the best; it hugs to the top left corner and has the maximum AUC value.*

3. **Gini coefficient**: We also use Gini coefficient to measure the goodness of a fit of our model. Formally put, it is the ratio of areas in a ROC curve and is a scaled version of the AUC.

$$GI = 2 * AUC - 1 \qquad \text{(Equation 3-7)}$$

   Similar to AUC values, a higher-value Gini coefficient is preferred.

4. **F1 score**: Many times, we face the problem of which KPI to choose (i.e., higher precision or higher recall) when we are comparing the models. F1 score solves this dilemma for us.

$$\text{F1 Score} = \frac{2(\text{precision} * \text{recall})}{\text{precision} + \text{recall}} \qquad \text{(Equation 3-8)}$$

   F1 score is the harmonic mean of precision and recall. The higher the F1 score, the better.

5. **AIC and BIC**: Akaike information criteria (AIC) and Bayesian information criteria (BIC) are used to choose the best model. AIC is derived from most frequent probability, while BIC is derived from Bayesian probability.

$$AIC = \frac{-2}{N} * LL + 2 * \frac{k}{N} \qquad \text{(Equation 3-9)}$$

while

$$BIC = -2 * LL + \log(N) * k \qquad \text{(Equation 3-10)}$$

In both formulas, N is number of examples in training set, LL is log-likelihood of the model on training dataset, and k is the number of variables in the model. The log in BIC is natural log to the base e and is called natural algorithm.

We prefer lower values of AIC and BIC. AIC penalizes the model for its complexity, but BIC penalizes the model more than AIC. If we have a choice between the two, AIC will choose a more complex model as compared to BIC.

Tip Given a choice between a very complex model and a simple model with comparable accuracy, choose the simpler one. Remember, nature always prefers simplicity!

---

6. **Concordance and discordance**: Concordance is one of the measures to gauge your model. Let us first understand the meaning of concordance and discordance.

   Consider if you are building a model to predict if a customer will churn from the business or not. The output is the probability of churn. The data is shown in Table 3-1.

***Table 3-1.** Respective Probability Scores for a Customer to Churn or Not*

Cust ID	Probability	Churn
1001	1	0.75
2001	0	0.24
3001	1	0.34
4001	0	0.62

Group 1: (churn = 1): Customer 1001 and 3001

Group 2: (churn = 0): Customer 2001 and 4001

Now we create the pairs by taking a single data point from Group 1 and one for Group 2 and then compare them. Hence, they will look like this:

Pair 1: 1001 and 2001

Pair 2: 1001 and 4001

Pair 3: 3001 and 2001

Pair 4: 3001 and 4001

By analyzing the pairs, we can easily make out that in the first three pairs the model is classifying the higher probability as churners. Here the model is correct in the classification. These pairs are called concordant pairs. Pair 4 is where the model is classifying the lower probability as churner, which does not make sense. This pair is called discordant. If the two pairs have comparable probabilities, this is referred to as tied pairs.

We can measure the quality using Somers D, which is given by

Somers D = (percentage concordant pair – percentage discordant pair). The higher the Somers D, the better the model.

Concordance alone cannot be a parameter to make a model selection. It should be used as one of the measures and other measures should also be checked.

7. **KS stats**: KS statistics or Kolmogorov-Smirnov statistics is one of the measures to gauge the efficacy of the model. It is the maximum difference between the cumulative true positive and cumulative false positive. The higher the KS, the better the model.

8. It is also a recommended practice to test the performance of the model on the following datasets and compare the KPIs:

    a. Training dataset: the dataset used for training the algorithm

    b. Testing dataset: the dataset used for testing the algorithm

    c. Validation dataset: this dataset is used only once and in the final validation stage

    d. Out-of-time validation: It is a good practice to have out-of-time testing. For example, if the training/testing/validation datasets are from Jan 2015 to Dec 2017, we can use Jan 2018–Dec 2018 as the out-of-time sample. The objective is to test a model's performance on an unseen dataset.

These are the various measures which are used to check the model's accuracy. We generally create more than one model using multiple algorithms. And for each algorithm, there are multiple iterations done. Hence, these measures are also used to compare the models and pick and choose the best one.

There is one more point we should be cognizant of. We would always want our systems to be accurate. We want to predict better if the share prices are going to increase or decrease or whether it will rain tomorrow or not. But sometimes, accuracy can be dubious. We will understand it with an example.

For example, while developing a credit card fraud transaction system our business goal is to detect transactions which are fraudulent. Now generally most (more than 99%) of the transactions are not fraudulent. This means that if a model predicts each incoming transaction as genuine, still the model will be 99% accurate! But the model is not meeting its business objective of spotting fraud transactions. In such a business case, recall is the important parameter we should target.

With this, we conclude our discussion of accuracy assessment. Generally, for a classification problem, logistic regression is the very first algorithm we use to baseline. Let us now solve an example of a logistic regression problem.

Case Study: Credit Risk

Business Context: Credit risk is nothing but the default in payment of any loan by the borrower. In the banking sector, this is an important factor to be considered before approving the loan of an applicant. Dream Housing Finance company deals in all home loans. They have presence across all urban, semiurban, and rural areas. Customers first apply for a home loan; after that company validates the customers' eligibility for the loan.

Business Objective: The company wants to automate the loan eligibility process (real time) based on customer detail provided while filling out the online application form. These details are gender, marital status, education, number of dependents, income, loan amount, credit history, and others. To automate this process, they have given a problem to identify the customer segments that are eligible for loan amounts so that they can specifically target these customers. Here they have provided a partial dataset.

Dataset: The dataset and the code is available at the Github link for the book shared at the start of the chapter. A description of the variables is given in the following:

Variable Description

a. Loan_ID: Unique Loan ID
b. Gender: Male/Female
c. Married: Applicant married (Y/N)
d. Dependents: Number of dependents
e. Education: Applicant Education (Graduate/Undergraduate)
f. Self_Employed: Self-employed (Y/N)
g. ApplicantIncome: Applicant income
h. CoapplicantIncome: Coapplicant income
i. LoanAmount: Loan amount in thousands
j. Loan_Amount_Term: Term of loan in months
k. Credit_History: credit history meets guidelines
l. Property_Area: Urban/ Semi Urban/ Rural
m. Loan_Status: Loan approved (Y/N)

Let's start the coding part using logistic regression. We will explore the dataset, clean and transform it, fit a model, and then measure the accuracy of the solution.

Step 1: Import all the requisite libraries first. Importing seaborn for statistical plots, to split data frames into training set and test set, we will use sklearn package's data-splitting function, which is based on random function. To calculate accuracy measures and confusion matrices, we have imported metrics from sklearn.

```
import pandas as pd
from sklearn.linear_model import LogisticRegression
import matplotlib.pyplot as plt
import seaborn as sns
from sklearn.model_selection import train_test_split
import numpy as np
import os,sys
from scipy import stats
from sklearn import metrics
import seaborn as sn
%matplotlib inline
```

Step 2: Load the dataset using read_csv command. The output is shown in the following.

```
loan_df = pd.read_csv('CreditRisk.csv')
loan_df.head()
```

	Loan_ID	Gender	Married	Dependents	Education	Self_Employed	ApplicantIncome	CoapplicantIncome	LoanAmount	Loan_Amount_Term	Cre
0	LP001002	Male	No	0	Graduate	No	5849	0.0	0	360.0	
1	LP001003	Male	Yes	1	Graduate	No	4583	1508.0	128	360.0	
2	LP001005	Male	Yes	0	Graduate	Yes	3000	0.0	66	360.0	
3	LP001006	Male	Yes	0	Not Graduate	No	2583	2358.0	120	360.0	
4	LP001008	Male	No	0	Graduate	No	6000	0.0	141	360.0	

Step 3: Examine the shape of the data:

```
loan_df.shape
```

Step 4: credit_df = loan_df.drop('Loan_ID', axis =1) # dropping this column as it will be 1-1 mapping anyways:

```
credit_df.head()
```

	Gender	Married	Dependents	Education	Self_Employed	ApplicantIncome	CoapplicantIncome	LoanAmount	Loan_Amount_Term	Credit_History
0	Male	No	0	Graduate	No	5849	0.0	0	360.0	1.0
1	Male	Yes	1	Graduate	No	4583	1508.0	128	360.0	1.0
2	Male	Yes	0	Graduate	Yes	3000	0.0	66	360.0	1.0
3	Male	Yes	0	Not Graduate	No	2583	2358.0	120	360.0	1.0
4	Male	No	0	Graduate	No	6000	0.0	141	360.0	1.0

Step 5: Next normalize the values of the Loan Value and visualize it too.

```
credit_df['Loan_Amount_Term'].value_counts(normalize=True)
plt.hist(credit_df['Loan_Amount_Term'], 50)
```

```
360.0    0.853333
180.0    0.073333
480.0    0.025000
300.0    0.021667
84.0     0.006667
240.0    0.006667
120.0    0.005000
36.0     0.003333
60.0     0.003333
12.0     0.001667
Name: Loan_Amount_Term, dtype: float64
```

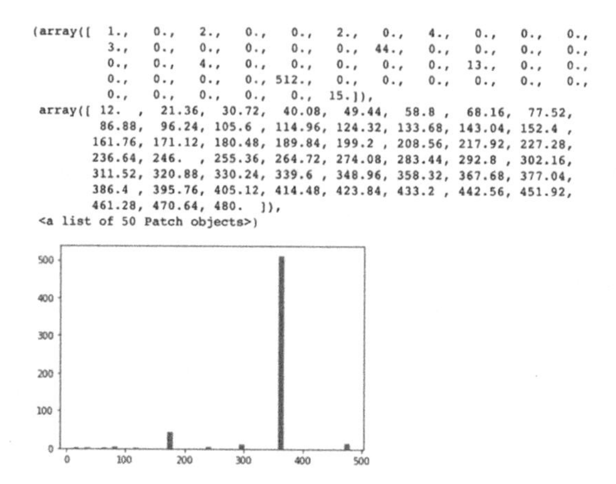

Step 6: Visualize the data next like a line chart.

```
plt.plot(credit_df.LoanAmount)
plt.xlabel('Loan Amount')
plt.ylabel('Frequency')
plt.title("Plot of the Loan Amount")
```

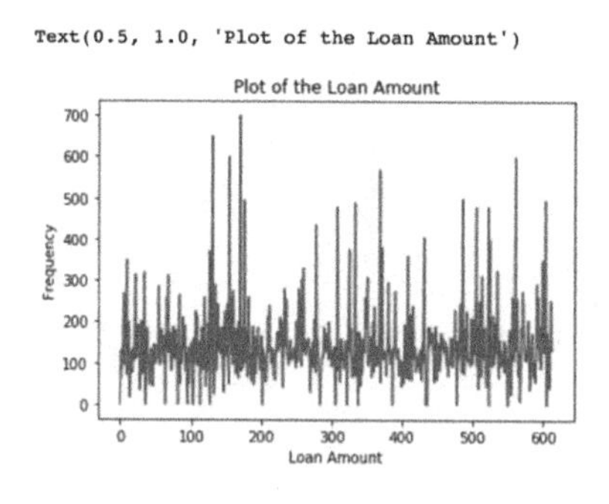

Tip We have shown only one visualization. You are advised to generate more graphs and plots. Remember, plots are a fantastic way to represent data intuitively!

Step 7: The Loan_Amount_Term is highly skewed and hence we are deleting this variable.

```
credit_df.drop(['Loan_Amount_Term'], axis=1, inplace=True)
```

Step 8: Missing value treatment is done next and each variable's missing value is replaced with 0. Compare the results after replacing the missing values with median.

```
credit_df = credit_df.fillna('0')
##credit_df = credit_df.replace({'NaN':credit_df.median()})
credit_df
```

Step 9: Next we will analyze how our variables are distributed.

```
credit_df.describe().transpose()
```

	count	mean	std	min	25%	50%	75%	max
ApplicantIncome	614.0	5403.459283	6109.041673	150.0	2877.5	3812.5	5795.00	81000.0
CoapplicantIncome	614.0	1621.245798	2926.248369	0.0	0.0	1188.5	2297.25	41667.0
LoanAmount	614.0	141.166124	88.340630	0.0	98.0	125.0	164.75	700.0
Loan_Status	614.0	0.687296	0.463973	0.0	0.0	1.0	1.00	1.0

You are advised to create box-plot diagrams as we have discussed in Chapter 2.

Step 10: Let us look at the target column, 'Loan_Status', to understand how the data is distributed among the various values.

```
credit_df.groupby(["Loan_Status"]).mean()
```

```
credit_df.groupby(["Loan_Status"]).mean()
```

Loan_Status	ApplicantIncome	CoapplicantIncome	LoanAmount
0	5446.078125	1877.807292	142.557292
1	5384.068720	1504.516398	140.533175

Step 11: Now we will convert X & Y variable to a categorical variable.

```
credit_df['Loan_Status'] = credit_df['Loan_Status'].
astype('category')
credit_df['Credit_History'] = credit_df['Credit_History'].
astype('category')
```

Step 12: Check the data types present in the data we have now as shown in the output:

```
credit_df.info()

<class 'pandas.core.frame.DataFrame'>
RangeIndex: 614 entries, 0 to 613
Data columns (total 11 columns):
Gender               614 non-null object
Married              614 non-null object
Dependents           614 non-null object
Education            614 non-null object
Self_Employed        614 non-null object
ApplicantIncome      614 non-null int64
CoapplicantIncome    614 non-null float64
LoanAmount           614 non-null int64
Credit_History       614 non-null category
Property_Area        614 non-null object
Loan_Status          614 non-null category
dtypes: category(2), float64(1), int64(2), object(6)
memory usage: 44.7+ KB
```

Step 13: Check how the data is balanced. We will get the following output.

```
prop_Y = credit_df['Loan_Status'].value_counts(normalize=True)
print(prop_Y)
```

```
1    0.687296
0    0.312704
Name: Loan_Status, dtype: float64
```

There seems to be a slight imbalance in the dataset as one class is 31.28% and the other is 68.72%.

Note While the dataset is not heavily imbalanced, we will also examine how to deal with data imbalance in Chapter 5.

Step 14: We will define the X and Y variables now.

```
X = credit_df.drop('Loan_Status', axis=1)
Y = credit_df[['Loan_Status']]
```

Step 15: Using one-hot encoding we will convert the categorical variables to numeric variables:

```
X = pd.get_dummies(X, drop_first=True)
```

Step 16: Now split into training and test sets. We are splitting into a ratio of 70:30

```
from sklearn.model_selection import train_test_split
X_train, X_test, y_train, y_test = train_test_split(X, Y, test_
size=0.30)
```

Step 17: Build the actual logistic regression model now:

```
import statsmodels.api as sm
logit = sm.Logit(y_train, sm.add_constant(X_train))
lg = logit.fit()
```

Step 18: We will now check the summary of the model. The results are as follows:

```
from scipy import stats
stats.chisqprob = lambda chisq, df: stats.chi2.sf(chisq, df)
print(lg.summary())
```

```
                           Logit Regression Results
==============================================================================
Dep. Variable:            Loan_Status   No. Observations:                  429
Model:                          Logit   Df Residuals:                      411
Method:                           MLE   Df Model:                           17
Date:                Sat, 30 May 2020   Pseudo R-squ.:                  0.2447
Time:                        18:47:23   Log-Likelihood:                -205.18
converged:                      False   LL-Null:                       -271.66
Covariance Type:            nonrobust   LLR p-value:                 5.042e-20
============================================================================================
                              coef    std err          z      P>|z|      [0.025      0.975]
--------------------------------------------------------------------------------------------
const                       9.2044    518.073      0.018      0.986   -1006.200    1024.609
ApplicantIncome         -3.429e-05      3e-05     -1.142      0.253    -9.31e-05    2.46e-05
CoapplicantIncome       -7.622e-05   4.03e-05     -1.891      0.059      -0.000    2.78e-06
LoanAmount                  0.0005      0.002      0.297      0.767      -0.003       0.004
Gender_Female               0.3004      0.885      0.340      0.734      -1.434       2.034
Gender_Male                 0.1726      0.826      0.209      0.835      -1.447       1.792
Married_No                -12.3041    518.072     -0.024      0.981   -1027.706    1003.098
Married_Yes               -11.9522    518.072     -0.023      0.982   -1027.354    1003.450
Dependents_1               -0.4619      0.346     -1.336      0.181      -1.139       0.216
Dependents_2                0.4379      0.405      1.082      0.279      -0.355       1.231
Dependents_3+               0.2402      0.533      0.450      0.652      -0.805       1.285
Education_Not Graduate     -0.2026      0.306     -0.661      0.508      -0.803       0.398
Self_Employed_No            0.2148      0.496      0.433      0.665      -0.758       1.187
Self_Employed_Yes           0.5411      0.604      0.895      0.371      -0.644       1.726
Credit_History_1.0          3.6165      0.497      7.276      0.000       2.642       4.591
Credit_History_0            3.6766      0.642      5.727      0.000       2.418       4.935
Property_Area_Semiurban     0.8764      0.318      2.757      0.006       0.253       1.500
Property_Area_Urban         0.0512      0.298      0.172      0.863      -0.532       0.634
============================================================================================
```

Let us interpret the results. The pseudo r-square shows that only 24% of the entire variation in the data is explained by the model. It is really not a good model!

Step 19: Next we will calculate the odds ratio from the coefficients using the formula odds ratio=exp(coef). Next we will calculate the probability from the odds ratio using the formula probability = odds / (1+odds).

```
log_coef = pd.DataFrame(lg.params, columns=['coef'])
log_coef.loc[:, "Odds_ratio"] = np.exp(log_coef.coef)
log_coef['probability'] = log_coef['Odds_ratio']/(1+log_
coef['Odds_ratio'])
log_coef['pval']=lg.pvalues
pd.options.display.float_format = '{:.2f}'.format
```

Step 20: We will now filter all the independent variables by significant p-value (p value <0.1) and sort descending by odds ratio. We will get the following output:

```
log_coef = log_coef.sort_values(by="Odds_ratio",
ascending=False)
pval_filter = log_coef['pval']<=0.1
log_coef[pval_filter]
```

	coef	Odds_ratio	probability	pval
Credit_History_0	3.68	39.51	0.98	0.00
Credit_History_1.0	3.62	37.21	0.97	0.00
Property_Area_Semiurban	0.88	2.40	0.71	0.01
CoapplicantIncome	-0.00	1.00	0.50	0.06

If we analyze the data, we can see that the customers who have credit history 1 have a 97% probability of defaulting the loan while the ones having history of 0 have 98% probability of defaulting.

Similarly, the customers in semiurban areas have odds of 2.50 times to default as compared to others.

Step 21: We are now fitting the model using the training data. The .fit is the function used for it.

```
from sklearn import metrics
from sklearn.linear_model import LogisticRegression
log_reg = LogisticRegression()
log_reg.fit(X_train, y_train)
```

Step 22: Once the model is ready and fit, we can use it to make a prediction. But first we have to check the accuracy of the model on the training data using confusion matrix; the output is as follows.

```
pred_train = log_reg.predict(X_train)
from sklearn.metrics import classification_report,
confusion_matrix
mat_train = confusion_matrix(y_train,pred_train)
print("confusion matrix = \n",mat_train)
```

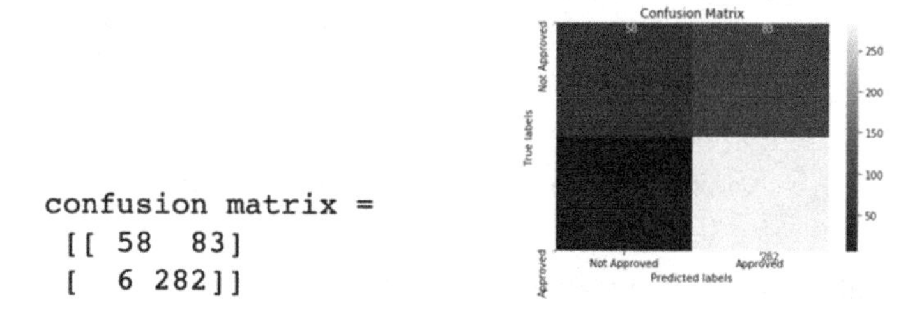

Step 23: Next we will make the prediction for test set and visualize it, and we will get the following output.

```
pred_test = log_reg.predict(X_test)
mat_test = confusion_matrix(y_test,pred_test)
print("confusion matrix = \n",mat_test)
ax= plt.subplot()
ax.set_ylim(2.0, 0)
annot_kws = {"ha": 'left',"va": 'top'}

sns.heatmap(mat_test, annot=True, ax = ax, fmt= 'g',
annot_kws=annot_kws); #annot=True to annotate cells
ax.set_xlabel('Predicted labels');
ax.set_ylabel('True labels');
ax.set_title('Confusion Matrix');
ax.xaxis.set_ticklabels(['Not Approved', 'Approved']);
ax.yaxis.set_ticklabels(['Not Approved', 'Approved']);
```

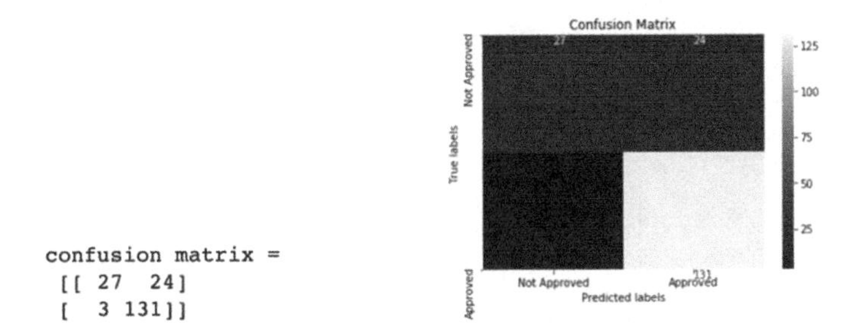

Step 24: Let us now create the AUC ROC curve and get the AUC score. We will get the following output.

```
from sklearn.metrics import roc_auc_score
from sklearn.metrics import roc_curve
logit_roc_auc = roc_auc_score(y_test, log_reg.predict(X_test))
fpr, tpr, thresholds = roc_curve(y_test, log_reg.predict_
proba(X_test)[:,1])
plt.figure()
plt.plot(fpr, tpr, label='Logistic Regression (area = %0.2f)' %
logit_roc_auc)
plt.plot([0, 1], [0, 1],'r--')
plt.xlim([0.0, 1.0])
plt.ylim([0.0, 1.05])
plt.xlabel('False Positive Rate')
plt.ylabel('True Positive Rate')
plt.title('Receiver operating characteristic')
plt.legend(loc="lower right")
plt.savefig('Log_ROC')
plt.show()
```

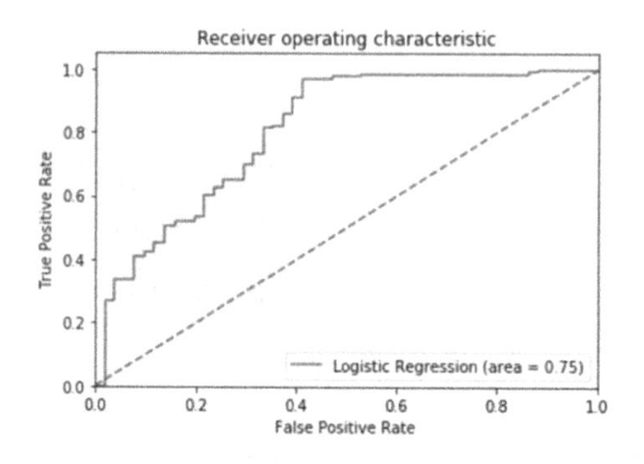


```
auc_score = metrics.roc_auc_score(y_test, log_reg.predict_
proba(X_test)[:,1])
round( float( auc_score ), 2 )
The output is 0.81.
```

Interpretations of the Results: By comparing the training confusion matrix and testing confusion matrix, we can determine the efficacy of the solution as shown in the confusion matrix.

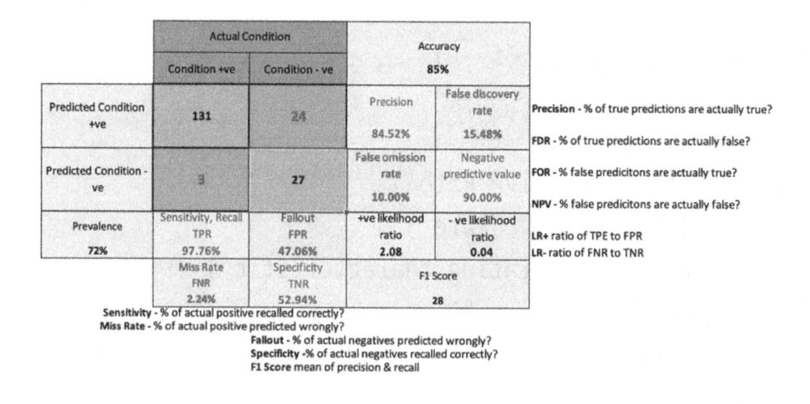

On testing data, the model's overall accuracy is 85%. Sensitivity or recall is 97% and precision is 84%. The model has a good overall accuracy. However, the model can be improved as we can see that 24 applications were predicted as approved while they were actually not approved.

Additional Notes

You can do a quick visualization for all the variables in using only one command using the following code and as shown in the following graph. It depicts the relationship of loan status with all the variables.

```
import seaborn as sns
sns.pairplot(credit_df, hue="Loan_Status", palette="husl")
```

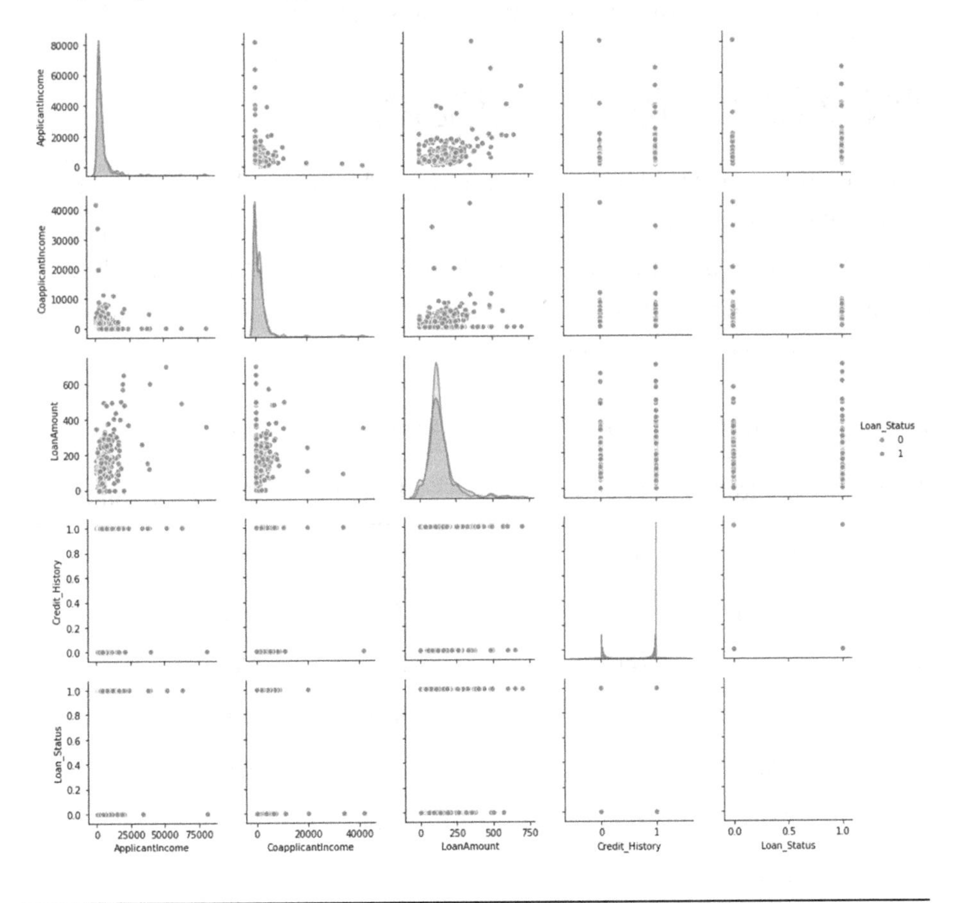

Note If the testing accuracy is not similar to training accuracy and is significantly lower, it means the model is overfitting. We will study how to tackle this problem in Chapter 5.

Logistic regression is generally the first few algorithms which are used whenever we approach a classification problem. It is fast, easy to comprehend, and compact to handle categorical and continuous data points alike. Hence, it is quite popular. It can also be used to get the significant variables for a problem.

Now that we have examined logistic regression in detail, let us move to the second very important classifier used: naïve Bayes. Don't go by the word "naïve"; this algorithm is quite robust in making classifications!

Naïve Bayes for Classification

Consider this: you are planning for camping. This trip will depend on a few factors like how the weather is, are there any predictions of rain, did you get a day off from the office, are your friends coming, and so on. You have the data from the history to make a prediction if you are going to camping or not which is shown in Table 3-2.

***Table 3-2.** Factors to Consider When Planning a Camping Trip*

Day off	Weather	Friends coming	Humidity	Going camping
Yes	Rainy	No	High	Yes
No	Sunny	Yes	Low	Yes
Yes	Overcast	No	Low	No
Yes	Rainy	No	High	Yes
Yes	Sunny	Yes	Low	Yes
Yes	Rainy	Yes	High	Yes
No	Sunny	No	High	No
No	Overcast	Yes	Low	Yes

As shown in the table, the final decision to go camping or not is dependent on the outcome from other events. This brings us to the concepts of *conditional probability*. We will first discuss a few key points related to probability to help understand better:

1) If A is any event, then the complement of A, denoted by $\hat{A}$, is the event that A does not occur.

2) The probability of A is represented by P(A), and the probability of its complement $P(\hat{A}) = 1 - P(A)$.

3) Let A and B be any events with probabilities P(A) and P(B). If you are told that when B has occurred, then the probability of A might change. As in the previous case if the weather is rainy then the probability of camping changes. This new probability of A is called the *conditional probability* of A given B, which can be written as P(A|B).

4) Mathematically, $P(A|B) = \frac{P(A \text{ and } B)}{P(B)}$ where P(A|B) means that probability of A given B which means the probability of A if B was known to have occurred.

5) This relationship can be viewed as *probabilistic* dependency and is called *conditional* probability. It means that knowledge of one event is of importance when assessing the probability of the other.

6) If the two events are mutually independent, then the multiplication rule simplifies to $P(A \text{ and } B) = P(A)P(B)$. For example, there will be no impact on your camping plans based on the price of milk.

There are many events which are mutually dependent on each other and hence it becomes imperative to understand the relationship: P(A|B) and P(B|A). This is true in the case of business activities too where the sales are dependent on the number of customers visiting the store, a customer will come back for shopping or not will depend on the previous experiences, and so on. Bayes' theorem helps to model for such factors and make a prediction.

As per *Bayes's rule,* if we have two events A and B. Then the conditional probability of A given B can be represented as

$$P(A|B) = P(B|A) \times \frac{P(A)}{P(B)} \qquad \text{(Equation 3-11)}$$

where P(A) and P(B): probability of A and B respectively, P(A|B): probability of A given B and P(B|A): probability of B given A.

For example, if we want to know in finding out a patient's probability to have a heart disease if they have diabetes. The data we have is as follows: 10% of patients entering the clinic have heart disease while 5% of the patients have diabetes. Among the patients diagnosed with heart disease, 8% are diabetic. Then P(A) = 0.10, P(B) = 0.05, and P(B|A) = 0.08. And using Bayes' rule, P(A|B) = (0.08×0.1)/0.05 = 0.16.

If we generalize the rule, let A_1 through A_n be a set of mutually exclusive outcomes. The probabilities of the events A are $P(A_1)$ through $P(A_n)$. These are called *prior* probabilities. Because an information outcome might influence our thinking about the probabilities of any A_i, we need to find the conditional probability $P(A_i|B)$ for each outcome A_i. This is called the *posterior* probability of A_i.

Using Bayes' rule, we can say that

$$P(A_i|B) = \frac{P(B|A_i)P(A_i)}{\{P(B|A_1)P(A_1) + P(B|A_2)P(A_2) + \ldots + P(B|A_n)P(A_n)}$$

(Equation 3-12)

Bayes' rule says that the posterior is the likelihood times the prior, divided by a sum of likelihood times priors. The denominator in Bayes' rule is the probability P(B).

$$\text{Posterior probability} = \frac{(\text{Conditional probability} \times \text{Prior probability})}{\text{evidence}}$$

(Equation 3-13)

Bayes' theorem is used for making classification (binary or multiclass), and it is referred to as *naïve Bayes.* It is called naïve due to a very strong assumption that the variables and features are independent of each other which is generally not true in the real world. Often this assumption is violated and still naïve Bayes tends to perform well. The idea is to factor all available evidence in the form of predictors into the naïve Bayes rule to obtain more accurate probability for class prediction.

As per Bayes' rule, the naïve Bayes estimates conditional probability (i.e., the probability that something will happen, given that something else has already occurred). For example, if we want to design an email spam filter and find a given mail is spam if there is an appearance of the word "discount." It is easy to implement, fast, robust, and quite accurate. Because of its ease of use, it is quite a popular technique.

Advantages of naïve Bayes algorithm:

1. It is a simple, easy, fast, and very robust method.
2. It does well with both clean and noisy data.
3. It requires few examples for training, but the underlying assumption is that the training dataset is a true representative of the population.
4. It is easy to get the probability for a prediction.

Disadvantages of naïve Bayes algorithm:

1. It relies on a very big assumption that independent variables are not related.
2. It is generally not suitable for datasets with large numbers of numerical attributes.
3. The predicted probabilities by the naïve Bayes algorithm is considered as less reliable in practice than predicted classes.
4. In some cases, it has been observed that if a rare event is not in training data but present in the test, then the estimated probability will be wrong.

But when some of our independent variables are continuous, we cannot calculate conditional probabilities! And hence in real-life variables, naïve Bayes is extended to Gaussian naïve Bayes.

In Gaussian naïve Bayes, continuous values associated with each attribute or independent variable are assumed to be following a Gaussian distribution. They are also easier to work with, as for the training we would only have to estimate the mean and standard deviation of the continuous variable.

Let us move to the case study to develop the naïve Bayes solution.

Case Study: Income Prediction on Census Data

Business Objective: We have census data and the objective is to predict whether income exceeds 50K/yr for an individual based on the value of other attributes.

Dataset: The dataset and code is available at the Github link shared at the start of this chapter.

Variable description:

- Age: continuous
- Workclass: Private, Self-emp-not-inc, Self-emp-inc, Federal-gov, Local-gov, State-gov, Without-pay, Never-worked.
- fnlwgt: continuous.
- Education: Bachelors, Some-college, 11th, HS-grad, Prof-school, Assoc-acdm, Assoc-voc, 9th, 7th-8th, 12th, Masters, 1st-4th, 10th, Doctorate, 5th-6th, Preschool.
- Education-num: continuous.
- Marital-status: Married-civ-spouse, Divorced, Never-married, Separated, Widowed, Married-spouse-absent, Married-AF-spouse.
- Occupation: Tech-support, Craft-repair, Other-service, Sales, Exec-managerial, Prof-specialty, Handlers-cleaners, Machine-op-inspct, Adm-clerical, Farming-fishing, Transport-moving, Priv-house-serv, Protective-serv, Armed-Forces.
- Relationship: Wife, Own-child, Husband, Not-in-family, Other-relative, Unmarried.
- Race: White, Asian-Pac-Islander, Amer-Indian-Eskimo, Other, Black.
- Sex: Female, Male.
- Capital-gain: continuous.
- Capital-loss: continuous.
- Hours-per-week: continuous.

- Native-country: United States, Cambodia, England, Puerto Rico, Canada, Germany, Outlying-US (Guam-USVI-etc), India, Japan, Greece, South, China, Cuba, Iran, Honduras, Philippines, Italy, Poland, Jamaica, Vietnam, Mexico, Portugal, Ireland, France, Dominican Republic, Laos, Ecuador, Taiwan, Haiti, Colombia, Hungary, Guatemala, Nicaragua, Scotland, Thailand, Yugoslavia, El Salvador, Trinadad&Tobago, Peru, Hong, Holland-Netherlands.
- Class: >50K, <=50K

Step 1: Import the necessary libraries here.

```
import pandas as pd
import numpy as np
from sklearn import preprocessing
from sklearn.model_selection import train_test_split
# used to split the dataset into train and test datasets
from sklearn.naive_bayes import GaussianNB
# To model the Gaussian Naive Bayes classifier
from sklearn.metrics import accuracy_score
# To calculate the accuracy score of the model
```

Step 2: We now import the data. Please note that this file has a .data extension. Now for importing the census data, we are passing four parameters. The 'adult.data' parameter is the file name. The header parameter suggests whether the first row of data consists of headers of the file or not. For our dataset there are no headers and hence we can pass 'None'. The delimiter parameter indicates the delimiter that is separating the data. Here, we are using the ' , ' delimiter. This delimiter allows the method to delete the spaces before and after the data values. This is very helpful when there is inconsistency in spaces used with data values.

```
census_df = pd.read_csv('adult.data', header = None,
delimiter=' *, *', engine='python')
```

Step 3: We are now adding the headers to the dataframe. It is required so that we are able to access the columns later and better:

```
census_df.columns = ['age', 'workclass', 'fnlwgt',
'education', 'education_num', 'marital_status', 'occupation',
'relationship', 'race', 'sex', 'capital_gain', 'capital_loss',
'hours_per_week', 'native_country', 'income']
```

Step 4: Let us print the total number of records (rows) in the dataframe:

```
len(census_df)
The output is 32561
```

Step 5: Check the presence of null values in our dataset, as follows:

```
census_df.isnull().sum()
```

```
age               0
workclass         0
fnlwgt            0
education         0
education_num     0
marital_status    0
occupation        0
relationship      0
race              0
sex               0
capital_gain      0
capital_loss      0
hours_per_week    0
native_country    0
income            0
dtype: int64
```

The preceding output shows that there is no "null" value in our dataset.

There can be some categorical variables having missing values. We will check that, sometimes they have "?" in place of missing values.

```
for value in ['workclass','education','marital_status','occupation',
'relationship','race','sex','native_country','income']:
    print(value,":", sum(census_df[value] == '?'))
```

```
workclass : 1836
education : 0
marital_status : 0
occupation : 1843
relationship : 0
race : 0
sex : 0
native_country : 583
income : 0
```

The output of the preceding code snippet shows that there are 1836 missing values in the workclass attribute, 1843 missing values in the occupation attribute, and 583 values in the native_country attribute.

Step 6: We will now proceed to data preprocessing. First, we will create a deep copy of our data frame:

```
census_df_rev = census_df.copy(deep=True)
```

Step 7: Before doing missing value handling tasks, we need some summary statistics of our data frame. For this, we can use the describe() method. It can be used to generate various summary statistics, excluding NaN values.

```
census_df_rev.describe(), as follows:
```

	age	fnlwgt	education_num	capital_gain	capital_loss	hours_per_week
count	32561.000000	3.256100e+04	32561.000000	32561.000000	32561.000000	32561.000000
mean	38.581647	1.897784e+05	10.080679	1077.648844	87.303830	40.437456
std	13.640433	1.055500e+05	2.572720	7385.292085	402.960219	12.347429
min	17.000000	1.228500e+04	1.000000	0.000000	0.000000	1.000000
25%	28.000000	1.178270e+05	9.000000	0.000000	0.000000	40.000000
50%	37.000000	1.783560e+05	10.000000	0.000000	0.000000	40.000000
75%	48.000000	2.370510e+05	12.000000	0.000000	0.000000	45.000000
max	90.000000	1.484705e+06	16.000000	99999.000000	4356.000000	99.000000

```
census_df_rev.describe(include= 'all')
```

If all is passed, it means we want to check the summary of all the attributes as follows:

	age	workclass	fnlwgt	education	education_num	marital_status	occupation	relationship	race	sex	capital_gain
count	32561.000000	32561	3.256100e+04	32561	32561.000000	32561	32561	32561	32561	32561	32561.000000
unique	NaN	9	NaN	16	NaN	7	15	6	5	2	NaN
top	NaN	Private	NaN	HS-grad	NaN	Married-civ-spouse	Prof-specialty	Husband	White	Male	NaN
freq	NaN	22696	NaN	10501	NaN	14976	4140	13193	27816	21790	NaN
mean	38.581647	NaN	1.897784e+05	NaN	10.080679	NaN	NaN	NaN	NaN	NaN	1077.648844
std	13.640433	NaN	1.055500e+05	NaN	2.572720	NaN	NaN	NaN	NaN	NaN	7385.292085
min	17.000000	NaN	1.228500e+04	NaN	1.000000	NaN	NaN	NaN	NaN	NaN	0.000000
25%	28.000000	NaN	1.178270e+05	NaN	9.000000	NaN	NaN	NaN	NaN	NaN	0.000000
50%	37.000000	NaN	1.783560e+05	NaN	10.000000	NaN	NaN	NaN	NaN	NaN	0.000000
75%	48.000000	NaN	2.370510e+05	NaN	12.000000	NaN	NaN	NaN	NaN	NaN	0.000000
max	90.000000	NaN	1.484705e+06	NaN	16.000000	NaN	NaN	NaN	NaN	NaN	99999.000000

Step 8: We will now impute the missing categorical values:

```
for value in ['workclass','education','marital_status','occupation',
'relationship','race','sex','native_country','income']:
    replaceValue = census_df_rev.describe(include='all')
    [value][2]
    census_df_rev[value][census_df_rev[value]=='?'] =
    replaceValue
```

Step 9: One-hot encoding to convert all the categorical variables to numeric

```
le = preprocessing.LabelEncoder()
workclass_category = le.fit_transform(census_df.workclass)
education_category = le.fit_transform(census_df.education)
marital_category   = le.fit_transform(census_df.marital_status)
occupation_category = le.fit_transform(census_df.occupation)
relationship_category = le.fit_transform(census_df.relationship)
race_category = le.fit_transform(census_df.race)
sex_category = le.fit_transform(census_df.sex)
native_country_category = le.fit_transform(census_df.native_country)
```

Step 10: We will now initialize the encoded categorical columns:

```
census_df_rev['workclass_category'] = workclass_category
census_df_rev['education_category'] = education_category
census_df_rev['marital_category'] = marital_category
census_df_rev['occupation_category'] = occupation_category
census_df_rev['relationship_category'] = relationship_category
census_df_rev['race_category'] = race_category
census_df_rev['sex_category'] = sex_category
census_df_rev['native_country_category'] = native_country_category
```

Step 11: Look at the first few lines of our data:

```
census_df_rev.head()
```

	age	workclass	fnlwgt	education	education_num	marital_status	occupation	relationship	race	sex	...	native_country
0	39	State-gov	77516	Bachelors	13	Never-married	Adm-clerical	Not-in-family	White	Male	...	United-States
1	50	Self-emp-not-inc	83311	Bachelors	13	Married-civ-spouse	Exec-managerial	Husband	White	Male	...	United-States
2	38	Private	215646	HS-grad	9	Divorced	Handlers-cleaners	Not-in-family	White	Male	...	United-States
3	53	Private	234721	11th	7	Married-civ-spouse	Handlers-cleaners	Husband	Black	Male	...	United-States
4	28	Private	338409	Bachelors	13	Married-civ-spouse	Prof-specialty	Wife	Black	Female	...	Cuba

5 rows × 23 columns

Step 12: There is no need of old categorical columns and we can drop them safely:

```
dummy_fields = ['workclass','education','marital_status',
'occupation','relationship','race', 'sex', 'native_country']
census_df_rev = census_df_rev.drop(dummy_fields, axis = 1)
```

Step 13: We will have to reindex all the columns and for that we will use the reindex_axis method.

```
census_df_rev = census_df_rev.reindex_axis(['age', 'workclass_
category', 'fnlwgt', 'education_category', 'education_num',
'marital_category', 'occupation_category', 'relationship_
category', 'race_category', 'sex_category', 'capital_gain',
'capital_loss', 'hours_per_week', 'native_country_category',
'income'], axis= 1) census_df_rev.head(5)
```

```
---------------------------------------------------------------------------
AttributeError                            Traceback (most recent call last)
<ipython-input-21-c1341a3f278e> in <module>
----> 1 census_df_rev = census_df_rev.reindex_axis(['age', 'workclass_category', 'fnlwgt', 'education_category',
      2                                   'education_num', 'marital_category', 'occupation_category',
      3                                   'relationship_category', 'race_category', 'sex_category', 'capital_gain',
      4                                   'capital_loss', 'hours_per_week', 'native_country_category',
      5                                   'income'], axis= 1)

~/opt/anaconda3/lib/python3.7/site-packages/pandas/core/generic.py in __getattr__(self, name)
   5177             if self._info_axis._can_hold_identifiers_and_holds_name(name):
   5178                 return self[name]
-> 5179             return object.__getattribute__(self, name)
   5180 
   5181     def __setattr__(self, name, value):

AttributeError: 'DataFrame' object has no attribute 'reindex_axis'
```

The method has deprecated and hence we will receive this error, as shown in the preceding illustration. Hence, let's use a newer method and we will not get the error shown.

```
census_df_rev = census_df_rev.reindex(['age', 'workclass_
category', 'fnlwgt', 'education_category', 'education_num',
'marital_category', 'occupation_category',  'relationship_
category', 'race_category', 'sex_category', 'capital_
gain',  'capital_loss', 'hours_per_week', 'native_country_
category', 'income'], axis= 1)
census_df_rev.head(5)
```

	age	workclass_category	fnlwgt	education_category	education_num	marital_category	occupation_category	relationship_category	race_category
0	39	7	77516	9	13	4	1	1	4
1	50	6	83311	9	13	2	4	0	4
2	38	4	215646	11	9	0	6	1	4
3	53	4	234721	1	7	2	6	0	2
4	28	4	338409	9	13	2	10	5	2

Step 14: We will now get our data arranged into dependent variables and target variable:

```
X = census_df_rev.values[:,:14]  ## These are the input variables
Y = census_df_rev.values[:,14]  ## This is the Target variable
```

Step 15: Now split the data into train and test in the ratio of 75:25.

```
X_train, X_test, Y_train, Y_test = train_test_split(X, Y,
test_size = 0.25, random_state = 5)
```

Step 16: We will fit the naïve Bayes model now.

```
clf = GaussianNB()
clf.fit(X_train, Y_train)
```

Step 17: The model classifier is now trained using training data and is ready to make predictions. We can use the predict() method with test set features as its parameters.

```
Y_pred = clf.predict(X_test)
```

Step 18: Check the accuracy of the model now.

```
accuracy_score(Y_test, Y_pred, normalize = True)
The accuracy we are getting is 0.79032.
```

With it we have implemented naïve Bayes using a live dataset. You are advised to understand each of the steps and practice the solution by replicating it.

Naïve Bayes is a fantastic algorithm to practice and use. Bayesian statistics is gathering a lot of attention and its power is being harnessed in research areas a lot. You might have heard the term *Bayesian optimization.* The beauty of the Bayes' theorem lies in its simplicity, which is very much visible in our day-to-day life. A straightforward method indeed!

We have covered logistic regression and naïve Bayes so far. Now we will examine one more very widely used classifier called *k-nearest neighbor.* One of the popular methods, easy to understand and implement—let's study knn next.

k-Nearest Neighbors for Classification

"Birds of the same feather flock together." This old adage is perfect for k-nearest neighbors. It is one of the most popular ML techniques where the learning is based on the similarity of data points with each other. knn is a *nonparametric* model; it does not construct a "model" and the classification is based on a simple majority vote from the neighbors. It can be used for classification where the relationship between attributes and target classes is complex and difficult to understand, and yet items in a class tend to be fairly homogenous on the values of attributes. But it might not be the best choice for an unclean dataset or where the target classes are not distinctively clear. If the target classes are not clearly demarcated, then it leads to an obvious confusion while taking the majority vote. knn can also be used for regression problems to make a prediction for a continuous variable. In the case of regression, the final output will be the average of the values of the neighbors and that average will be assigned to the target variable. Let us examine this visually:

For example, we have some data points represented by circles and plus signs in a vector space diagram as shown in Figure 3-5. There are clearly two classes in this case. The objective is to classify a new data point (marked in yellow) shown in Figure 3-5(ii) and identify which class it belongs to.

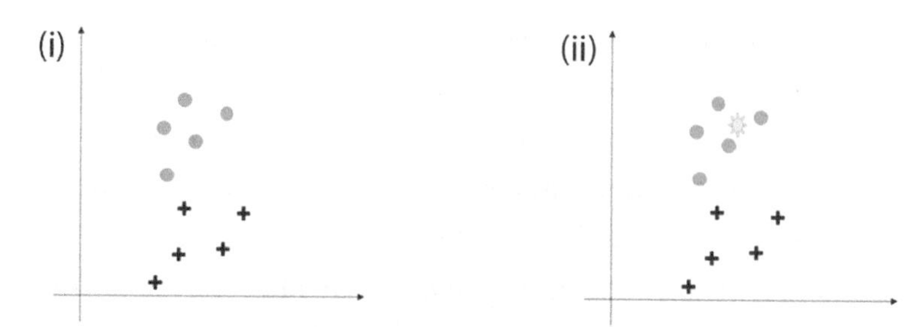

Figure 3-5. *(i) The distribution of two classes in green circle and black plus sign. (ii) The right side shows the new data point which has to be classified (shown in yellow sign). The k-nearest neighbor algorithm has to be used to make the classification.*

This yellow point can be a circle or a plus sign and nothing else. knn will help in this classification by taking a vote of majority from the other data points in the vicinity. And the value of "k" will guide us on how many data points to be considered for voting. Let's assume we took k = 4. Hence, we will now make a circle with a yellow point as the center. And the circle should be just as big as to enclose only four data points. It is represented in Figure 3-6(i).

The four closest points to the yellow point all belong to circles or we can say that all the *neighbors* of the unknown yellow point are circles. Hence, with a good confidence level we can predict that the yellow point should belong to the circle. Here the choice was comparatively easy and straightforward. Refer to Figure 3-6(ii), where it is not that simple. Hence, the choice of k plays a very crucial role.

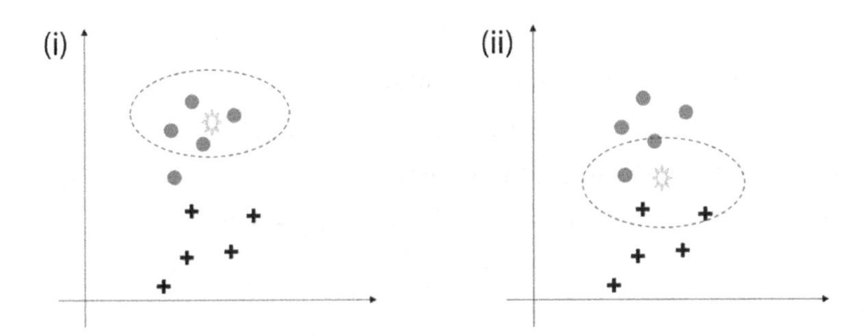

Figure 3-6. *(i) The presence of new unseen data if we select four nearest neighbors. It is an easy decision to make. (ii) The right side shows that the new data point is difficult to classify as the four neighbors are mixed.*

The steps which are followed in k-nearest neighbor are as follows:

1. We receive the raw and unclassified dataset which has to be worked upon.

2. We choose a distance matrix from Euclidean, Manhattan or Minkowski.

3. Then calculate the distance between the new data points and the known classified training data points.

4. The number of neighbors to be considered is defined by the value of "k".

5. It is followed by comparing with the list of classes which have the shortest distance and count the number of times each class appears.

6. The class with the highest votes wins. This means that the class which has the highest frequency and has appeared the greatest number of times is assigned to the unknown data point.

Tip Parametric models make some assumptions about the input data like having a normal distribution. However, nonparametric methodology believes that data distributions are undefinable by a finite set of parameters and hence do not make any assumptions.

From the steps discussed for k-nearest neighbor, we can clearly understand that the final accuracy depends on the distance matrix used and the value of "k".

Popular distance matrices used are

1. **Euclidean Distance**: probably the most common and easiest way to calculate between two points. It is square root of the sum of the squares of distances:

$$\text{Euclidean Distance} = \sqrt{\left((y_2 - y_1)^2 + (x_2 - x_1)^2\right)} \qquad \text{(Equation 3-14)}$$

2. **Manhattan Distance**: The distance between two points measured along axes at right angles. Sometimes it is also referred to as city block distance. In a plane with p_1 at (x_1, y_1) and p_2 at (x_2, y_2), it is

$$\text{Manhattan Distance} = |x_1 - x_2| + |y_1 - y_2| \qquad \text{(Equation 3-15)}$$

3. **Minkowski Distance**: This is a metric in a normed vector space. Minkowski distance is used for distance similarity of vectors. Given two or more vectors, find the distance similarity of these vectors. Mainly, the Minkowski distance is applied in ML to

find out the distance similarity. It is a generalized distance metric and can be represented by the following formula:

$$\left(\Sigma_{i=1}^{n}\left|x_i - y_i\right|^p\right)^{1/p} \qquad \text{(Equation 3-16)}$$

where by using different values of p we can get different values of distances. With the value of p = 1 we get Manhattan distance, with p = 2 we get Euclidean distance, and with p= ∞ we get Chebychev distance.

4. **Cosine Similarity**: It is a measure of similarity between two nonzero vectors of an inner product space that measures the cosine of the angle between them. The cosine of 0° is 1, and it is less than 1 for any angle in the interval (0,π] radians.

The various distances can be viewed as shown in Figure 3-7.

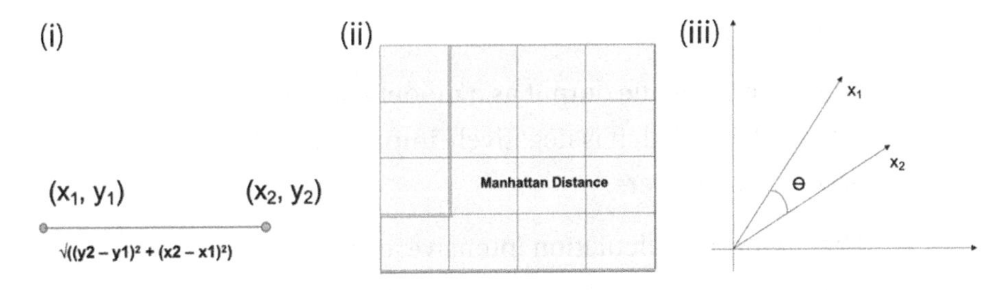

Figure 3-7. *(i) Euclidean distance; (ii) Manhattan distance; (iii) cosine similarity*

Tip While approaching a knn problem, generally we start with Euclidean distance. In most business problems, it serves the purpose. Distance matrix is an important parameter in unsupervised clustering methods like k-means clustering.

Advantages of k-nearest neighbor

1. It is a nonparametric method and does not make any assumptions about distributions of the various classes in vector space.

2. It can be used for the binary classification as well as multiclassification problems.

3. The method is quite easy to comprehend and implement.

4. The method is robust and if the value of k is large, it is not impacted by outliers.

Disadvantages of k-nearest neighbor

1. The accuracy depends on the value of k and hence finding the most optimal value can be a difficulty sometimes.

2. The method requires the class distributions to be non-overlapping.

3. There is no specific output as a model and if the value of k is small, it is negatively impacted by the presence of outliers.

4. The method is calculation intensive as it is a *lazy learner*. The distances have to be calculated between all the points and then a majority is to be taken. And hence, it is not that fast of a method to use.

There are other forms of knn too, which are as follows:

Radius Neighbor Classifier

1. This classifier implements the learning based on a number of neighbors. The neighbors are within a fixed radius r of each training point, where r is a floating point value specified by the user.

2. We prefer this method when the data sampling is not uniform. But, in case of quite a few independent variables and a sparse dataset, it suffers with the *curse of dimensionality.*

Tip When the number of dimensions increases, the volume of space increases at a very fast pace and the resultant data becomes very sparse. This is called the curse of dimensionality. Data sparsity makes statistical analysis for any dataset quite a challenging task.

KD Tree Nearest Neighbor

1. This method is effective if the dataset is large but the number of independent variables is less.

2. The method takes less time to compute as compared to other methods.

It is now time to create a Python solution using knn, which we are executing next.

Case Study: k-Nearest Neighbor

The dataset to be audited, which consists of a wide variety of intrusions simulated in a military network environment, was provided. It created an environment to acquire raw TCP/IP dump data for a network by simulating a typical US Air Force LAN. The LAN was focused like a real environment and blasted with multiple attacks. A connection is a sequence of TCP packets starting and ending at some time duration between which data flows to and from a source IP address to a target IP address under some well-defined protocol. Also, each connection is labeled either as normal or as an attack with exactly one specific attack type. Each connection record consists of about 100 bytes. For each TCP/IP connection, 41 quantitative and qualitative features are obtained from normal and attack data (3 qualitative and 38 quantitative features).

The class variable has two categories: normal and anomalous.

The Dataset

The dataset is available at the git repository as `Network_Intrusion.csv` file. The code is also available at the Github link shared at the start of the chapter.

Business Objective

We have to fit a k-nearest neighbor algorithm to detect network intrusion.

Step 1: Import all the required libraries. We are importing `pandas, numpy, matplotlib, seaborn.`

```
import pandas as pd
import numpy as np
import seaborn as sns
import matplotlib.pyplot as plt
%matplotlib inline
```

Step 2: Import the dataset using read.csv method from pandas. Let's have a look at the top five rows of the data first as follows:

```
network_data= pd.read_csv('Network_Intrusion.csv')
network_data.head()
```

	duration	protocol_type	service	flag	src_bytes	dst_bytes	land	wrong_fragment	urgent	hot	...	dst_host_srv_count	dst_host_same_srv_rate
0	0	1	19	9	491	0	0	0	0	0	...	25	0.17
1	0	2	41	9	146	0	0	0	0	0	...	1	0.00
2	0	1	46	5	0	0	0	0	0	0	...	26	0.10
3	0	1	22	9	232	8153	0	0	0	0	...	255	1.00
4	0	1	22	9	199	420	0	0	0	0	...	255	1.00

5 rows × 42 columns

Step 3: Now we will do the regular checkup of the data using info() and describe command.

```
network_data.info()
```

```
network_data.info()

<class 'pandas.core.frame.DataFrame'>
RangeIndex: 25192 entries, 0 to 25191
Data columns (total 42 columns):
duration                          25192 non-null int64
protocol_type                     25192 non-null int64
service                           25192 non-null int64
flag                              25192 non-null int64
src_bytes                         25192 non-null int64
dst_bytes                         25192 non-null int64
land                              25192 non-null int64
wrong_fragment                    25192 non-null int64
urgent                            25192 non-null int64
hot                               25192 non-null int64
num_failed_logins                 25192 non-null int64
logged_in                         25192 non-null int64
num_compromised                   25192 non-null int64
root_shell                        25192 non-null int64
su_attempted                      25192 non-null int64
num_root                          25192 non-null int64
```

```
network_data.describe().transpose()
```

```
network_data.describe().transpose()
```

	count	mean	std	min	25%	50%	75%	max
duration	25192.0	305.054104	2.686556e+03	0.0	0.00	0.00	0.00	42862.0
protocol_type	25192.0	1.053827	4.269982e-01	0.0	1.00	1.00	1.00	2.0
service	25192.0	29.039139	1.555560e+01	0.0	19.00	22.00	46.00	65.0
flag	25192.0	6.982455	2.679322e+00	0.0	5.00	9.00	9.00	10.0
src_bytes	25192.0	24330.628215	2.410805e+06	0.0	0.00	44.00	279.00	381709090.0
dst_bytes	25192.0	3491.847174	8.883072e+04	0.0	0.00	0.00	530.25	5151385.0
land	25192.0	0.000079	8.909946e-03	0.0	0.00	0.00	0.00	1.0
wrong_fragment	25192.0	0.023738	2.602208e-01	0.0	0.00	0.00	0.00	3.0
urgent	25192.0	0.000040	6.300408e-03	0.0	0.00	0.00	0.00	1.0
hot	25192.0	0.198039	2.154202e+00	0.0	0.00	0.00	0.00	77.0
num_failed_logins	25192.0	0.001191	4.541818e-02	0.0	0.00	0.00	0.00	4.0
logged_in	25192.0	0.394768	4.888105e-01	0.0	0.00	0.00	1.00	1.0

Step 4: Now check for the null values. In our dataset there are no null values fortunately.

```
network_data.isnull().sum()
```

Note This dataset does not have any null values; we will study in detail on how to deal with null values, NA, NaN, and so on in Chapter 5.

Step 5: Let's have a look at the class distribution. And we will visualize it too.

```
network_data["class"].value_counts(normalize=True)
```

```
network_data["class"].value_counts(normalize=True)

1    0.53386
0    0.46614
Name: class, dtype: float64
```

```
pd.value_counts(network_data["class"]).plot(kind="bar")
```

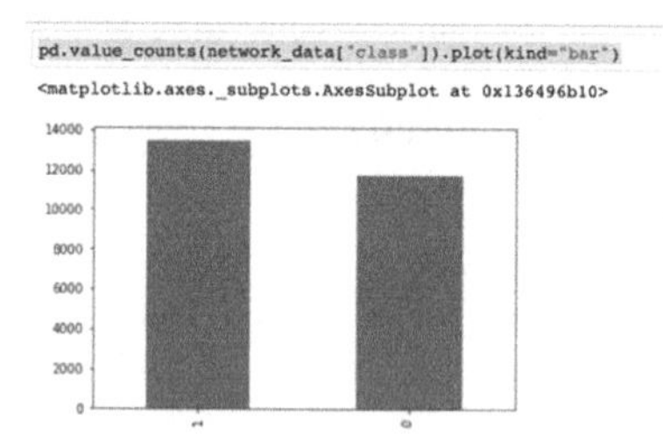

Step 6: There are a few categorical variables in our dataset. We have to convert them to numerical variables using *one-hot encoding*.

```
from sklearn.preprocessing import LabelEncoder
label_encoder = LabelEncoder()
network_data['class'] = label_encoder.
fit_transform(dataset['class'])
```

```
network_data['protocol_type'] = label_encoder.
fit_transform(dataset['protocol_type'])
network_data['service'] = label_encoder.
fit_transform(dataset['service'])
network_data['flag'] = label_encoder.fit_transform(dataset['flag'])
```

Step 7: One-hot encoding increases the number of variables in the dataset. Let's see the number of columns in the dataset with added variables:

```
network_data.columns
```

Step 8: Next we will standardize our dataset by using StandardScaler in scikit learn.

```
from sklearn import preprocessing
from sklearn.preprocessing import StandardScaler
X_std = pd.DataFrame(StandardScaler().fit_transform
(network_data))
X_std.columns = network_data.columns
```

Step 9: Now it is the time to split the data in train and test. We are dividing the data in an 80:20 ratio.

```
import numpy as np
from sklearn.cross_validation import train_test_split
X = np.array(network_data.ix[:, 1:5]) #Transform data into features
y = np.array(network_data['class']) #Transform data into targets
X_train, X_test, y_train, y_test = train_test_split(X, y,
test_size=0.2, random_state=7)
```

```
# split into train and test
X_train, X_test, y_train, y_test = train_test_split(X, y, test_size=0.2, random_state=7)
---------------------------------------------------------------------------
ModuleNotFoundError                       Traceback (most recent call last)
<ipython-input-57-3ff897371412> in <module>
      2 import numpy as np
      3
----> 4 from sklearn.cross_validation import train_test_split
      5
      6 # Transform data into features and target

ModuleNotFoundError: No module named 'sklearn.cross_validation'
```

Step 10: The sklearn.cross_validation has deprecated and hence we received this error. Again, try to split in train and test using sklearn.model_selection.

```
from sklearn.model_selection import train_test_split
# Transform data into features and target
X = np.array(network_data.ix[:, 1:5])
y = np.array(network_data['class'])
# split into train and test
X_train, X_test, y_train, y_test = train_test_split(X, y,
test_size=0.2, random_state=7)
```

Step 11: Print the shape of the data by print(X_train.shape).

```
print(y_train.shape)
```

```
print(X_train.shape)
print(y_train.shape)

(20153, 4)
(20153,)
```

Step 12: Print the shape of the test data by print(X_test.shape).

```
print(y_test.shape)
```

```
print(X_test.shape)
print(y_test.shape)

(5039, 4)
(5039,)
```

Step 13: We will now train the model using training data and iterate with different values of k=3,5,9.

```
from sklearn.neighbors import KNeighborsClassifier
from sklearn.metrics import accuracy_score
from sklearn.metrics import recall_score
```

```
# instantiate learning model (k = 3)
knn_model = KNeighborsClassifier(n_neighbors = 3)
Fitting the model
knn_model.fit(X_train, y_train)
y_pred = knn_model.predict(X_test) # predict the response
print(accuracy_score(y_test, y_pred)) # Evaluate accuracy
The answer is 0.9902758483826156
knn_model = KNeighborsClassifier(n_neighbors=5) # With k = 5
knn_model.fit(X_train, y_train) # Fitting the model
y_pred = knn_model.predict(X_test) # Predict the response
print(accuracy_score(y_test, y_pred)) # Evaluate accuracy
The answer is 0.9882913276443739
With k = 9
knn_model = KNeighborsClassifier(n_neighbors=9)
knn_model.fit(X_train, y_train) # Fitting the model
y_pred = knn_model.predict(X_test) # Predict the response
print(accuracy_score(y_test, y_pred)) # Evaluate accuracy
The answer is 0.9867037110537805
```

Step 14: We have tested with three values of k. We will now iterate on multiple values of k. We will run the knn with the no. of neighbors to be 1,3,5...19 and then find the optimal number of neighbors based on the lowest misclassification error.

```
k_list = list(range(1,20)) # creating odd list of K for KNN
k_neighbors = list(filter(lambda x: x % 2 != 0, k_list))
# subsetting just the odd ones
ac_scores = [] # empty list that will hold accuracy scores
# perform accuracy metrics for values from 1,3,5....19
for k in k_neighbors:
```

```
    knn_model = KNeighborsClassifier(n_neighbors=k)
    knn_model.fit(X_train, y_train)
    y_pred = knn_model.predict(X_test)   # predict the response
    scores = accuracy_score(y_test, y_pred)  # evaluate accuracy
    ac_scores.append(scores)
# changing to misclassification error
MSE = [1 - x for x in ac_scores]
# determining best k
optimal_k = k_neighbors[MSE.index(min(MSE))]
print("The optimal number of neighbors is %d" % optimal_k)
```

Step 15: Let's print the impact of different values of k on the misclassification error.

```
import matplotlib.pyplot as plt
# plot misclassification error vs k
plt.plot(k_neighbors, MSE)
plt.xlabel('Number of Neighbors K')
plt.ylabel('Misclassification Error')
plt.show()
```

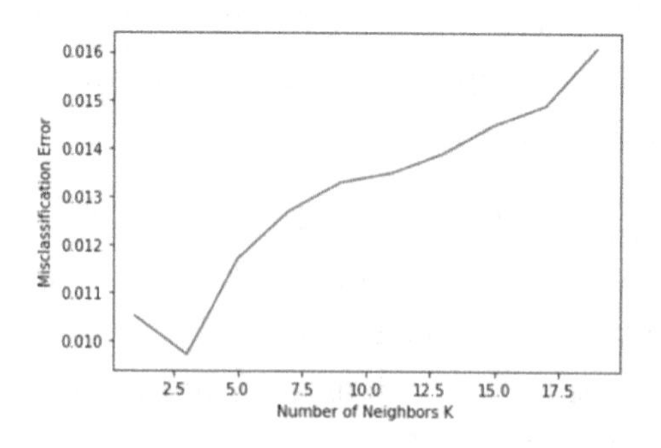

Step 16: It turns out that k=3 gives us the best result. Let's implement it:

```
#Use k=3 as the final model for prediction
knn = KNeighborsClassifier(n_neighbors = 3)
# fitting the model
knn.fit(X_train, y_train)
# predict the response
y_pred = knn.predict(X_test)
# evaluate accuracy
print(accuracy_score(y_test, y_pred))
print(recall_score(y_test, y_pred))
```

The accuracy and recall are 0.9902758483826156 and 0.9911944869831547 respectively.

With k = 3, we are getting a very good accuracy and a great recall value too. This model can be considered as the final one to be used.

We developed a solution using knn and are getting good accuracies. k-nearest neighbor is very easy to explain and visualize. Not being very statistics- or mathematics-heavy, even for non-data science users, the method is not very tedious to understand. And this is one of the most important properties of this method. Being a nonparametric method, there is no assumption about the data and hence it requires less data preparation.

We have covered logistic regression, naïve Bayes, and k-nearest neighbor. Generally, when we start any classification solution, we start with these three algorithms to check their accuracy. The next in the series are tree-based algorithms: decision tree and random forest. We have discussed the theoretical concepts about both in Chapter 2. In the next section, we will cover the differences and then implement them on a dataset.

Tree-Based Algorithms for Classification

Recall in Chapter 2 that we studied decision trees to predict the values of a continuous variable. Since decision trees can be used for both classification and regression problems, in this chapter we will study classification solutions using decision tree. The building blocks for a decision tree remain the same as shown in Figure 3-8.

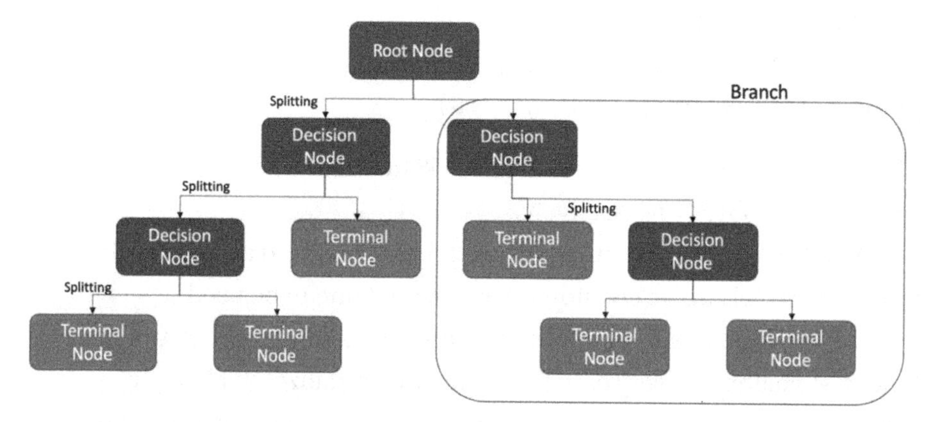

Figure 3-8. *A decision tree is comprised of root node, decision node, and a terminal node. A subtree is called a branch*

The difference is the process of splitting followed by a classification algorithm which is discussed now.

The objective of splitting is to create as many pure nodes as possible. If a resultant node after splitting contains all the data points belonging to the same class it is called *pure* or *homogeneous*. If the node contains records belonging to different classes, the node is *impure* or *heterogenous*. The objective is to create pure nodes. There are three primary ways to measure the impurity: entropy, Gini index, and classification error. We describe them in detail now and compare their respective processes. Consider that we have a dataset like in Table 3-3.

Table 3-3. *Transportation Mode Dependency on Other Factors Like Gender and Income Level*

Vehicle Count	Gender	Cost	Income	Transportation Mode
1	Female	Less	Middle class	Train
0	Male	Less	Low income	Bus
1	Female	High	Middle class	Train
1	Male	Less	Middle class	Bus
0	Male	Medium	Middle class	Train
1	Male	Less	Middle class	Bus
2	Female	High	High class	Car
0	Female	Less	Low income	Bus
2	Male	High	Middle class	Car
1	Female	High	High class	Car

We want to train an algorithm to predict the transportation mode. As per the preceding example, we can calculate that the respective probabilities are

```
Probability (Bus) = 4/10 = 0.4
Probability (Car) = 3/10 = 0.3
Probability (Train) = 3/10 = 0.3
```

The three methods are entropy, Gini index, and classification error, which are described now:

Entropy: *Entropy* and *information gain* walk hand-in-hand. A pure node will require less information to describe itself while an impure node will require more information. It can be understood in the form of entropy too. (Information gain = 1 - Entropy.)

$$\text{Entropy of the system} = -p * \log_2 p - q * \log_2 q \qquad \text{(Equation 3-17)}$$

where p and q are the probability of success and failure, respectively, in that node. The logarithm is to the base of 2 here.

In the preceding example, entropy = -0.4 (log 0.4) - 0.3(log 0.3) - 0.3(log 0.3) = 1.571

Entropy of a pure node is 0 and can be represented as Figure 3-9.

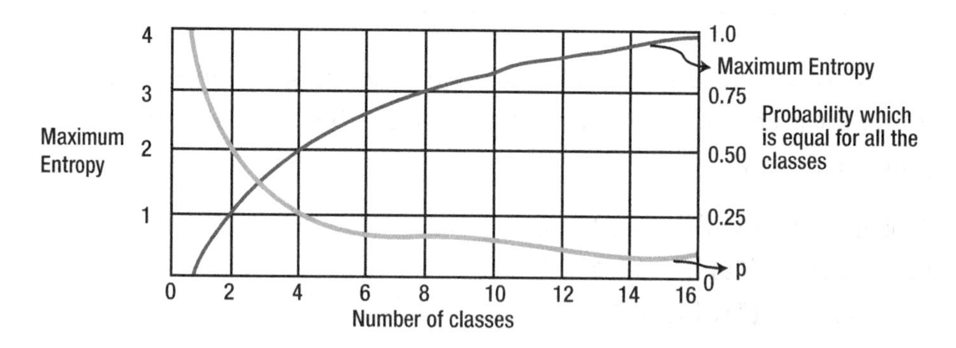

Figure 3-9. *Values of maximum entropy for different numbers of classes (n). Probability p = 1/n.*

Gini coefficient: Gini index can also be used to measure the impurity. The formula to be used is as follows:

$$\text{Gini Index} = 1 - \Sigma p^2 j \qquad \text{(Equation 3-18)}$$

In the preceding example, Gini index = 1 - (0.4^2 + 0.3^2 + 0.3^2) = 0.660

Gini of a node containing a single class is 0 because the probability is 1. Like entropy, it also takes maximum value when all the classes in the node have the same probability. The movement of Gini can be represented as in Figure 3-10.

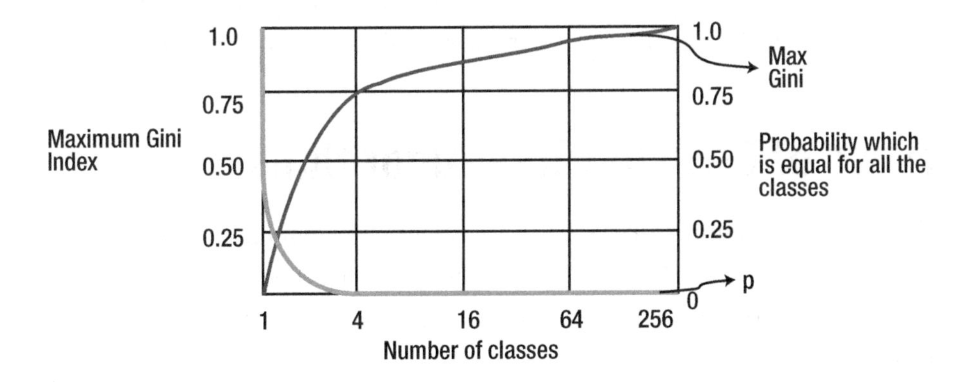

Figure 3-10. *Values of maximum Gini index for different number of classes (n). Probability p = 1/n.*

The value of Gini will always be between 0 and 1 irrespective of the number of classes in the model.

Classification Error: The next way to measure the degree of impurity is using classification error. It is given by the formula

$$\text{Classification Error Index} = 1 - \max(p_i) \qquad \text{(Equation 3-19)}$$

where i is the number of classes

Similar to the other two, its value lies between 0 and 1.

For the preceding example, classification error index = 1 - max(0.4,0.3,0.3) = 0.6

We can use any of these three splitting methods for classification. There are some common decision tree algorithms which are used for regression and classification problems. Since we have studied both regression and classification concepts, it is a good time to examine different types of decision tree algorithms, which is the next section.

Types of Decision Tree Algorithms

There are some significant decision tree algorithms which are used in the industry. Some of them are suitable for classification problems while some are a better choice for regression solutions. We are discussing all the aspects of the algorithms.

The prominent algorithms are as follows:

1. **ID3** or Iterative Dichotomizer 3 is a decision tree algorithm using greedy search to split the dataset in each iteration. It uses entropy or information gain as a factor to perform the split iteratively. For each successive iteration in the model, it uses unused variables in the last iteration, calculates the entropy for those unused variables, and then selects the

variable with lowest entropy. Or in other words, it selects the variable with highest information gain. ID3 can lead to overfitting and may not be the most optimal choice. It fares quite well with categorical variables. But when the dataset contains continuous variables, it becomes slower to converge since there are many values on which the node splitting can be done.

2. **CART** or classification and regression tree is a flexible tree-based solution. It can model for both continuous or categorical target variables and hence it is one of the highly used tree-based algorithms. Like a regular decision tree algorithm, we choose input variables and split the nodes iteratively till we achieve a robust tree to work upon. The selection of the input variables is done using a greedy approach with the objective to minimize the loss. The tree construction stops based on a predefined criteria like minimum observations to be present in each of the leaves. Python library scikit-learn uses an optimized version of CART but does not support categorical variables as of now.

3. **C4.5** is an extension of ID3 and is used for classification problems. Similar to ID3, it utilizes entropy or information gain to make the split. It is a robust choice since it can handle both categorical and continuous variables. For continuous variables, it assigns a threshold value and does the split based on the threshold. Variables with value above the threshold are in one bucket while variables with

threshold less than or equal to threshold are in a different bucket. It allows missing variables in the data as the missing values are not considered while calculating the entropy values.

4. **CHAID (Chi-square automatic interaction detection)** is a popular algorithm in the field of market research and marketing; for example, if we want to understand how a certain group of customers will respond to a new marketing campaign. This marketing campaign can be for a new product or service and will be useful for the marketing team to strategize accordingly. CHAID is primarily based on adjusted significance testing and is mostly used when we have a categorical target variable and categorical independent variables. It proves to be quite a handy and convenient method to visualize such a dataset.

5. **MARS** or multivariate adaptive regression splines is a nonparametric regression technique. It is mostly suitable for measuring nonlinear relationships between variables. It is a flexible regression model which can handle both categorical and continuous variables. It is quite a robust solution to handle massive datasets and requires very much less data preparation, making is comparatively faster to implement. Owing to its flexibility and ability to model nonlinearities in the dataset, MARS generally is a good choice to tackle overfitting in the model.

The tree-based algorithms discussed previously are unique in their own way. Some of them are more suitable for classification problems, while some are a better choice for regression problems. CART can be used for both classification and regression problems.

Now it is time to develop a case study using decision tree.

The code and the dataset are available at the Github link shared at the start of this chapter.

We can use the same dataset we have used for the logistic regression problem. The implementation follows after we have created the training and testing data.

Step 1: Import the necessary libraries first.

```
from sklearn.tree import DecisionTreeClassifier
```

Step 2: Now we are calling the decision tree classifier and training the model.

```
dt_classifier = DecisionTreeClassifier()
dt_classifier.fit(X_train, y_train)
```

Step 3: Use the trained model to make a prediction on the test data.

```
y_pred = dt_classifier.predict(X_test)
```

Step 4: Get the confusion matrix. To visualize it, you are advised to use the method used in the logistic regression method.

```
print(confusion_matrix(y_test, y_pred))
ax= plt.subplot()
ax.set_ylim(2.0, 0)
annot_kws = {"ha": 'left',"va": 'top'}
sns.heatmap(mat_test, annot=True, ax = ax, fmt= 'g',
annot_kws=annot_kws); #annot=True to annotate cells
```

```
ax.set_xlabel('Predicted labels');
ax.set_ylabel('True labels');
ax.set_title('Confusion Matrix');
ax.xaxis.set_ticklabels(['Not Approved', 'Approved']);
ax.yaxis.set_ticklabels(['Not Approved', 'Approved']);
```

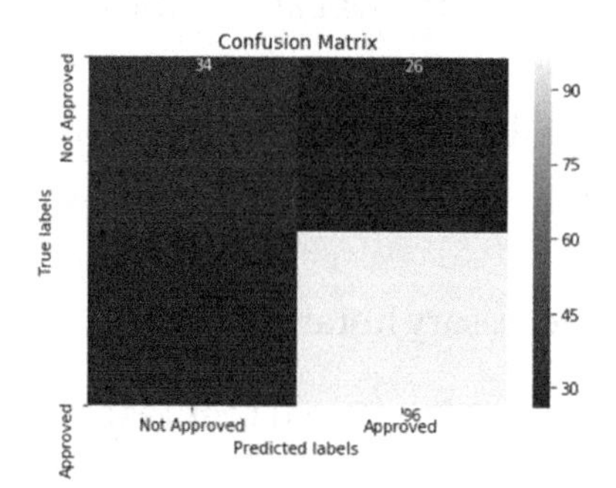

We will now model the same problem using a random forest model. Recall that random forest is an ensemble-based technique where it creates multiple smaller trees using a subset of data. The final decision is based on the *voting* mechanism by each of the trees. In the last chapter, we have used random forest for a regression problem; here we are using random forest for a classification problem.

Tip Decision trees are generally prone to overfitting; ensemble-based random forest model is a good choice to tackle overfitting.

Step 1: Import the library and fit the model. Create the model with 500 trees.

```
from sklearn.ensemble import RandomForestClassifier
rf_model = RandomForestClassifier(n_estimators=500, bootstrap = True,
                                  max_features = 'sqrt')
Now we will fit on training data
rf_model.fit(X_train, y_train)
```

Step 2: Predict on test data and plot the confusion matrix.

```
y_pred = rf_model.predict(X_test)
print(confusion_matrix(y_test, y_pred))
ax= plt.subplot()
ax.set_ylim(2.0, 0)
annot_kws = {"ha": 'left',"va": 'top'}
sns.heatmap(mat_test, annot=True, ax = ax, fmt= 'g',
annot_kws=annot_kws); #annot=True to annotate cells
ax.set_xlabel('Predicted labels');
ax.set_ylabel('True labels');
ax.set_title('Confusion Matrix');
ax.xaxis.set_ticklabels(['Not Approved', 'Approved']);
ax.yaxis.set_ticklabels(['Not Approved', 'Approved']);
```

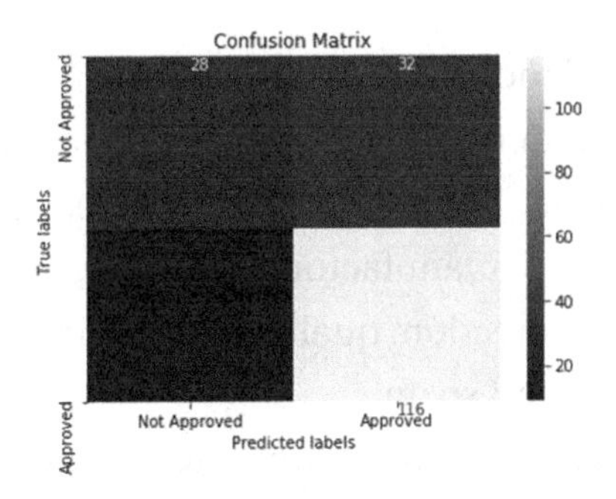

This is the implementation of a decision tree algorithm and ensemble-based random forest algorithm. Tree-based algorithms are very easy to comprehend and implement. They are generally the first few algorithms which we implement and test the accuracy of the system. Tree-based solutions are recommended if we want to create a quick solution, but they are prone to overfitting. We can use tree pruning or setting a constraint on tree size to overcome overfitting. We will again visit this concept in Chapter 5, in which we will discuss all the techniques to overcome the problem of overfitting in our ML model.

With this, we come to the end of our discussion on tree-based algorithms. In this chapter, we have studied the classification algorithms and implemented them too. These algorithms are quite popular in the industry and powerful enough to help us make a robust ML model. Generally, we test the data on these algorithms at the start and then choose the one which is giving us the best results. And then we tune it further till we achieve the most desirable output. The desirable output may *not* be the most complex solution but surely will be one which will deliver the desired level of measurement parameters, reproducibility, robustness, flexibility, and ease of deployment. Remember, complexity is *not* proportional to accuracy. A more complex model does *not* mean a higher degree of performance!

Summary

Prediction is a powerful tool in our hands. Using these ML algorithms, we can not only take a confident decision we can also ascertain the factors which affect that decision. These algorithms are heavily used across sectors like banking, retail, manufacturing, insurance, aviation, and so on. The uses include fraud detection, quality inspection, churn prediction, loan default prediction, and so on.

You should note that these algorithms are not the only sources of knowledge. A sound exploratory analysis is a prerequisite for a good ML algorithm. And the most important resource is "data" itself. A good-quality and representative dataset is of paramount importance. We have discussed the qualities of a good dataset in Chapter 1.

It is also imperative that a sound business problem has been created from the start. The choice of the target variable should align with the business problem at hand. The training data used to train the algorithm plays a very crucial role, as on it depends the patterns learned by the algorithm. It is important to note that we do measure the performance of the algorithms using the various parameters like precision, recall, AUC,

F1 score, and so on. An algorithm which is performing well on training, testing, and validation datasets will be the best algorithm. But still there are a few other parameters based on which we choose the final algorithm which can be deployed into production, which we discuss in Chapter 5.

In Chapter 1, we examined ML, various types, data and attributes of data quality and ML process. In Chapter 2, we studied ML algorithms to model a continuous variable. In this third chapter, we complemented the knowledge with classification algorithms. These first chapters have created a firm base for you to solve most of the business problems in the data science world. Also, you are now ready to take the next step in Chapter 4.

In the first three chapters, we have discussed basic and intermediate algorithms. In the next chapter, we are going to cover much more complex algorithms like SVM, gradient boosting, and neural networks for regression and classification problems. So stay focused!

EXERCISE QUESTIONS

Question 1: How does a logistic regression algorithm make a classification prediction?

Question 2: What is the difference between precision and recall?

Question 3: What is posterior probability?

Question 4: What are the assumptions in a naïve Bayes algorithm?

Question 5: How can we choose the value of k in k-nearest neighbor?

Question 6: What are the various performance measurement parameters for classification algorithms?

Question 7: The sinking of the ship *Titanic* in 1912 was indeed heart-breaking. Some passengers survived, some did not. Download the dataset from `https://www.kaggle.com/c/titanic`. Using ML, predict which passengers were more likely to survive than others based on the various attributes of the passengers.

Question 8: Load the dataset Iris using the following command:

```
from sklearn.datasets import load_iris
iris = load_iris()
```

We have worked upon the same dataset in the last chapter. Here, you have to classify the type of the flower using classification algorithms and compare the results.

Question 9: Download the Bank Marketing Dataset from the link `https://archive.ics.uci.edu/ml/datasets/Bank+Marketing`. The data is related with direct marketing campaigns (phone calls) of a Portuguese banking institution. The classification problem goal is to predict if the client will subscribe to a term deposit. Create various classification models and compare the respective KPIs.

Question 10: Get the German Credit Risk data from `https://www.kaggle.com/uciml/german-credit`. The dataset contains attributes of each person who takes credit from the bank. The objective is to classify each person as a good or bad credit risk according to their attributes. Use classification algorithms to create the model and choose the best model.

Question 11: Go through the research paper on logistic regression at `https://www.ncbi.nlm.nih.gov/pmc/articles/PMC6696525/`.

Question 12: Go through the research paper on random forest at `https://aip.scitation.org/doi/pdf/10.1063/1.4977376`. There is one more good paper at `https://aip.scitation.org/doi/10.1063/1.4952607`.

Question 13: Examine the research paper on knn at `https://pdfs.semanticscholar.org/a196/39771e987588b378879c65300b61b4af86af.pdf`.

Question 14: Study the research paper on naïve Bayes at `https://www.cc.gatech.edu/~isbell/reading/papers/Rish.pdf`.

CHAPTER 4

Advanced Algorithms for Supervised Learning

"A real intelligence is an art to simplify complex matters without losing the integrity of that matter."

— Sumit Singh

Our lives are complex. We have to deal with complexity everyday—at home, at work, with our commute, within family, and with our career goals. There are many paths to success, but the definition of success is subjective and complex. And we always strive to find the best ingredients to ease that path to success.

Like our lives too, data can be hugely complex at times. We need more advanced algorithms, sophisticated techniques, out-of-the-box approaches, and innovative processes to make sense of it. But the heart of any solution, any algorithm, any approach, and any process is the need to resolve the business problem at hand. Most of the business problems involve how to increase profits and to decrease costs. Using such advanced methodologies, we can make sense with the complex datasets that are generated by our systems.

V. Verdhan, *Supervised Learning with Python*,
https://doi.org/10.1007/978-1-4842-6156-9_4

In the first three chapters of the book, we studied ML for regression and classification problems using quite a few algorithms. We examined the concepts and developed Python solutions for them. In this chapter, we are going to work on advanced algorithms. We will be studying these algorithms and developing the mathematical concepts and coding logic for them. We will not be working on structured data alone. We will be working on unstructured datasets too—text and image—in this chapter.

In this chapter, advanced algorithms like boosting and SVMs will be examined. Then we will be diving into the world of text and image data, and solving such challenges using principles of natural language processing (NLP) and image analysis. Deep learning is used for solving the complex problems and hence we will be implementing deep learning to a structured data and unstructured image dataset. All the code files and datasets are provided with step-by-step explanations.

Technical Toolkit Required

We are going to use Python 3.5 or above in this book. You are advised to get Python installed on your machine. We will be using Jupyter notebook; installing Anaconda-Navigator is required for executing the codes. All the datasets and codes have been uploaded to the Github library at `https://github.com/Apress/supervised-learning-w-python/tree/master/Chapter%204` for easy download and execution.

The major libraries used are numpy, pandas, matplotlib, seaborn, scikit learn, and so on. You are advised to install these libraries in your Python environment. In this chapter, we are going to use NLP so will use NLTK library and RegexpTokenizer. We will also need Keras and TensorFlow libraries in this chapter.

Let us go into the ensemble-based boosting algorithms and study the concepts in detail!

Boosting Algorithms

Recall in the last chapter we studied ensemble-modeling techniques. We discussed bagging algorithms and created solutions using random forest. We will continue with ensemble-modeling techniques. The next algorithm is boosting.

Formally put, boosting is an ensemble method that creates a strong classifier from weak classifiers. In a sequence, we create a new model while assuring we are learning from the errors or misclassifications from the previous model as shown in Figure 4-1. The idea is to give higher importance to the errors and improve the modeling subsequently, finally resulting in a very strong model.

Initially, a subset is taken from the training dataset and all the data points are given equal weight. We create a base version of the model (let's call it M1) and then loss is calculated based on the wrong predictions. In the next iteration, the incorrect data points are awarded higher weights and another model (let's call it M2) is created. The idea is M2 will be better than M1 as it is improving and trying to correct the errors from M1, and this process continues where multiple models are created, each improving the previous one. The final model is the weighted mean of all the previous models. The final model acts as a strong learner.

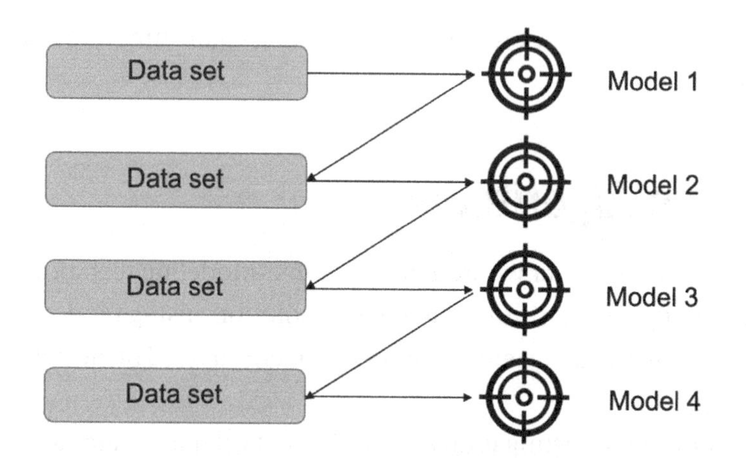

***Figure 4-1.** Boosting algorithms iteratively work and improve the previous version while assigning higher weights to the errors*

There are multiple types of boosting algorithms available:

1. **Gradient boosting**: Gradient boosting can work for both classification and regression problems. For example, a regression tree can be used as a base learner and each subsequent tree will be an improvement over the previous tree. The overall learner gradually improves on the observations where the residuals have been initially high.

   The properties of gradient boosting are as follows:

   a. A base learner is created by taking a subset of the complete dataset.

   b. The difficult observations are identified, or the *shifting* is done by identifying the value of the residuals in the previous model.

   c. The misclassifications are identified by the gradients calculated. This is the central idea of the algorithm. It creates new base learners which are correlated maximum with the negative gradient of the loss function, which in turn is associated with the complete ensemble solution.

d. It then further dissects error components to add more information about the residuals.

2. **AdaBoosting**: AdaBoosting or adaptive boosting is considered as a special case of gradient boosting, wherein iterative models are created to improve upon the previous model. Initially, a base model is created using a subset of the data and is used to make predictions on the complete dataset. We measure the performance by calculating the error. Then while creating the next model, data points which have been predicted incorrectly are given higher weights. The weights are proportional to the error; that is, the higher the error, the higher is the weight assigned. Hence, the next model created is an improvement over the previous model, and this process continues. Once it is no longer possible to reduce the error further, the process stops and we conclude that we have reached the final model, which is the best model.

   AdaBoost has the following properties:

   a. In AdaBoost, the shifting is done by assigning higher weights to the observations misclassified in the previous step.

   b. The misclassifications are identified by high-weight observations.

   c. The exponential loss in AdaBoost assigns greater value of weight for the sample which have ill-fitted in the previous model.

3. **Extreme gradient boosting**: Extreme gradient boosting of XGB is an advanced boosting algorithm. It has become quite popular lately and has won many data science and ML competitions. It is extremely accurate and quite a fast solution to implement.

   The properties for XGB are as follows:

   a. XGB is quite a fast algorithm since it allows parallel processing and hence is faster than standard gradient boosting.

   b. It tackles overfitting by implementing regularization techniques.

   c. It works well with messy datasets having missing values, as it has an inbuilt mechanism to handle missing values present in the dataset. This is one of the biggest advantages, as we do not have to deal with missing values present in the data.

   d. It is quite a flexible algorithm and allows us to have a customized optimization objective and evaluation criteria.

   e. Cross-validation at each iteration results in an optimum number of boosting iterations, which makes it a better choice than its counterparts.

4. **CatBoost**: CatBoost is a fantastic solution if we are dealing with categorical variables. In typical ML models, we use one-hot encoding to deal with

categorical variables. For example, if we have a dataset having a categorical variable as "City," we convert it to numeric variables as shown in Table 4-1.

Table 4-1. *One-Hot Encoding to Convert Categorical Variables to Numeric Variables*

CustID	Revenue	City	Items
1001	100	New Delhi	4
1002	101	London	5
1003	102	Tokyo	6
1004	104	New Delhi	8
1001	100	New York	4
1005	105	London	5

CustID	Revenue	New Delhi	London	Tokyo	New York	Items
1001	100	1	0	0	0	4
1002	101	0	1	0	0	5
1003	102	0	0	1	0	6
1004	104	1	0	0	0	8
1001	100	0	0	0	1	4
1005	105	0	1	0	0	5

But if we have 100 unique values for the variable "City," one-hot encoding will result in adding 100 additional dimensions to the dataset. Moreover, the resultant dataset will be quite *sparse*. Sparsity means that for a column only a few rows will be 1; the rest will be 0. For example, in Table 4-1 Tokyo has got only one value as 1. That means that the matrix contains more 0's than 1's. Hence, the performance operation across will take a long time. Moreover, if the number of resultant dimensions are too large then we will have huge memory requirements.

CatBoost does not suffer from this problem. CatBoost deals with categorical variables internally and we do not have to spend time on dealing with them.

5. **Light gradient boosting**: As the name suggests, light gradient boosting is computationally less expensive than its counterparts. It is the choice of boosting algorithm if the dataset is extremely large. It implements tree-based algorithms and uses a leaf-based approach, as compared to others, which use a level-based approach, as shown in Figure 4-2.

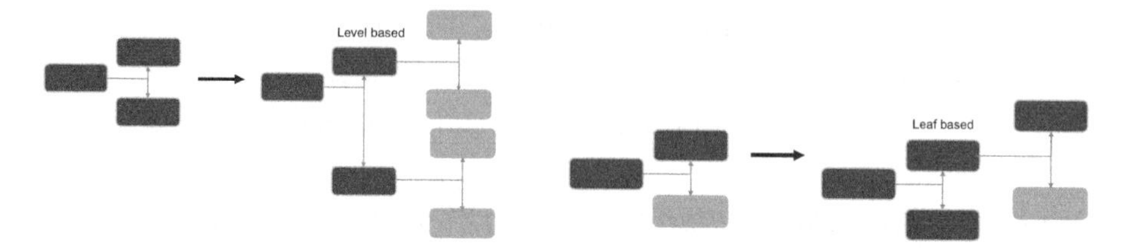

Figure 4-2. *Level-based approach is used by other boosting algorithms, while leaf-based approach makes light gradient boosting a good fit for large datasets*

We have now discussed the different types of boosting algorithms. Depending on the business problem at hand and the data set available, we will prefer one method over another. Recently extreme gradient boosting or XGB has gained a lot of popularity. It is quite a robust technique, gives better results, and deals with overfitting internally.

We will now implement a case in Python using gradient boosting algorithm.

Using Gradient Boosting Algorithm

In this case study, we are going to implement multiple algorithms. We have studied multiple algorithms till now, and some of them are ensemble-based advanced algorithms. It is the correct time to compare their respective accuracies.

We will perform EDA, create train-test split, and then implement decision tree, random forest, bagging, AdaBoost, and gradient boosting algorithm. Finally, we will compare the respective performance of all the algorithms.

The dataset and code can be downloaded from the Github link shared at the start of the chapter. The data is for predicting wine quality based on parameters like fixed acidity, volatile acidity, and so on.

Step 1: Import all the libraries first:

```
%matplotlib inline
import numpy as np
import pandas as pd
from sklearn.tree import DecisionTreeClassifier
import numpy as np
import pandas as pd
import seaborn as sns
from matplotlib import pyplot as plt
from sklearn.model_selection import train_test_split
from sklearn.tree import DecisionTreeClassifier
from sklearn import metrics
from sklearn.metrics import accuracy_score,f1_score,recall_
score,precision_score, confusion_matrix
%matplotlib inline

from sklearn.feature_extraction.text import CountVectorizer
```

Step 2: Import the dataset:

```
wine_quality_data_frame = pd.read_csv('winequality-red-1.
csv',sep=';')
```

Step 3: Print the first five samples from the data:

```
wine_quality_data_frame.head(5)
```

```
wine_quality_data_frame.head(5)
```

	fixed acidity	volatile acidity	citric acid	residual sugar	chlorides	free sulfur dioxide	total sulfur dioxide	density	pH	sulphates	alcohol	quality
0	7.4	0.70	0.00	1.9	0.076	11.0	34.0	0.9978	3.51	0.56	9.4	5
1	7.8	0.88	0.00	2.6	0.098	25.0	67.0	0.9968	3.20	0.68	9.8	5
2	7.8	0.76	0.04	2.3	0.092	15.0	54.0	0.9970	3.26	0.65	9.8	5
3	11.2	0.28	0.56	1.9	0.075	17.0	60.0	0.9980	3.16	0.58	9.8	6
4	7.4	0.70	0.00	1.9	0.076	11.0	34.0	0.9978	3.51	0.56	9.4	5

Step 4: Get the information about the data types:

```
wine_quality_data_frame.info()
```

```
wine_quality_data_frame.info()

<class 'pandas.core.frame.DataFrame'>
RangeIndex: 1599 entries, 0 to 1598
Data columns (total 12 columns):
fixed acidity           1599 non-null float64
volatile acidity        1599 non-null float64
citric acid             1599 non-null float64
residual sugar          1599 non-null float64
chlorides               1599 non-null float64
free sulfur dioxide     1599 non-null float64
total sulfur dioxide    1599 non-null float64
density                 1599 non-null float64
pH                      1599 non-null float64
sulphates               1599 non-null float64
alcohol                 1599 non-null float64
quality                 1599 non-null int64
dtypes: float64(11), int64(1)
memory usage: 150.0 KB
```

Step 5: Get the details about all the numeric variables present in the dataset:

```
wine_quality_data_frame.describe()
```

```
wine_quality_data_frame.describe()
```

	fixed acidity	volatile acidity	citric acid	residual sugar	chlorides	free sulfur dioxide	total sulfur dioxide	density	pH	sulphates	alcohol	
count	1599.000000	1599.000000	1599.000000	1599.000000	1599.000000	1599.000000	1599.000000	1599.000000	1599.000000	1599.000000	1599.000000	1[illegible]
mean	8.319637	0.527821	0.270976	2.538806	0.087467	15.874922	46.467792	0.996747	3.311113	0.658149	10.422983	
std	1.741096	0.179060	0.194801	1.409928	0.047065	10.460157	32.895324	0.001887	0.154386	0.169507	1.065668	
min	4.600000	0.120000	0.000000	0.900000	0.012000	1.000000	6.000000	0.990070	2.740000	0.330000	8.400000	
25%	7.100000	0.390000	0.090000	1.900000	0.070000	7.000000	22.000000	0.995600	3.210000	0.550000	9.500000	
50%	7.900000	0.520000	0.260000	2.200000	0.079000	14.000000	38.000000	0.996750	3.310000	0.620000	10.200000	
75%	9.200000	0.640000	0.420000	2.600000	0.090000	21.000000	62.000000	0.997835	3.400000	0.730000	11.100000	
max	15.900000	1.580000	1.000000	15.500000	0.611000	72.000000	289.000000	1.003690	4.010000	2.000000	14.900000	

Step 6: We will perform some analysis and visualizations on the dataset now:

```
import matplotlib.pyplot as plt
import seaborn as sns
sns.countplot(wine_quality_data_frame['quality'])
```

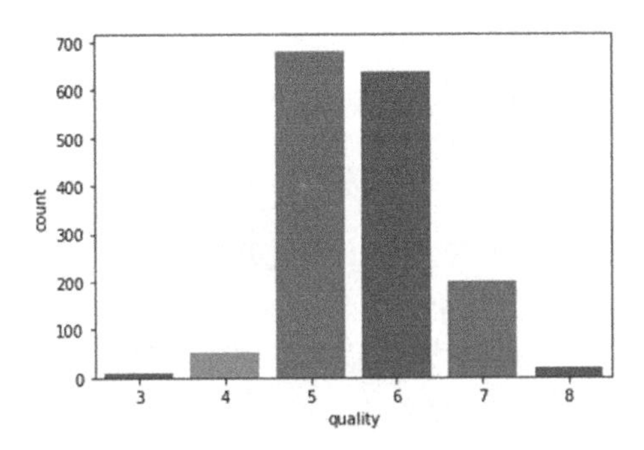


```
sns.distplot(wine_quality_data_frame['volatile acidity'])
```

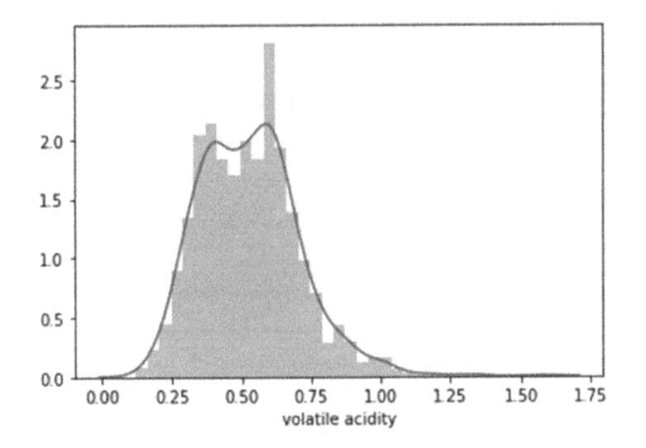

Correlation plot will be created to observe the relationships between the variables:

```
plt.figure(figsize=(10,10))
sns.heatmap(wine_quality_data_frame.corr(),
            annot=True,
            linewidths=.5,
            center=0,
            cbar=False,
            cmap="Blues")
plt.show()
```

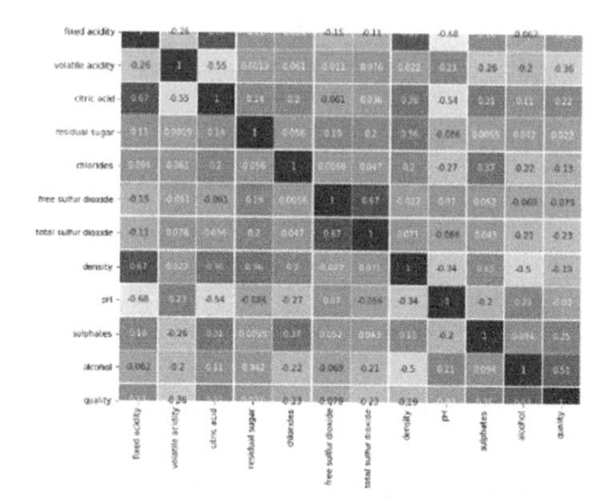

Step 7: We will analyze the frequency of target variables.

```
wine_quality_data_frame['quality'].value_counts()
```

```
5     681
6     638
7     199
4      53
8      18
3      10
Name: quality, dtype: int64
```

Step 8: We are combining a few levels here so that we can have a balanced target variable. As we can observe, levels 3, 4, and 8 are having lower values hence combining with other levels.

```
wine_quality_data_frame['quality'] = wine_quality_data_
frame['quality'].replace(8,7)
wine_quality_data_frame['quality'] = wine_quality_data_
frame['quality'].replace(3,5)
wine_quality_data_frame['quality'] = wine_quality_data_
frame['quality'].replace(4,5)
wine_quality_data_frame['quality'].value_counts()
```

Step 9: Now split into training and testing data:

```
from sklearn.model_selection import train_test_split

X_train, X_test, y_train, y_test =train_test_split(wine_
quality_data_frame.drop('quality',axis=1), wine_quality_data_
frame['quality'], test_size=.20, random_state=5)
X_train.shape,X_test.shape
```

Step 10: Decision tree implementation is done next:

```
dt_entropy=DecisionTreeClassifier(criterion='entropy')
dt_entropy.fit(X_train, y_train)
dt_entropy.score(X_train, y_train)
dt_entropy.score(X_test, y_test)
```

The training accuracy is 100% and testing is 69% which means that the model is overfitting. Hence, we are pruning the tree to maximum depth of 4

```
clf_pruned = DecisionTreeClassifier(criterion = "entropy",
random_state = 50, max_depth=4, min_samples_leaf=6)
clf_pruned.fit(X_train, y_train)

inde_variables = wine_quality_data_frame.drop('quality', axis=1)
feature_column = inde_variables.columns
prediction_pruned = clf_pruned.predict(X_test)
prediction_pruned_train = clf_pruned.predict(X_train)
print(accuracy_score(y_test,prediction_pruned))
print(accuracy_score(y_train,prediction_pruned_train))
acc_DT = accuracy_score(y_test, prediction_pruned)
```

Step 11: The overfitting has been handled but accuracy has not improved.

We are now getting the significant features for our dataset.

```
feature_importance = clf_pruned.tree_.compute_feature_
importances(normalize=False)
```

```
feat_imp_dict = dict(zip(feature_column, clf_pruned.feature_
importances_))
feat_imp = pd.DataFrame.from_dict(feat_imp_dict,
orient='index')
feat_imp.sort_values(by=0, ascending=False)
```

	0
alcohol	0.481629
sulphates	0.266467
volatile acidity	0.100496
fixed acidity	0.089209
free sulfur dioxide	0.036360
total sulfur dioxide	0.025839
citric acid	0.000000
residual sugar	0.000000
chlorides	0.000000
density	0.000000
pH	0.000000

Step 12: We can deduce that alcohol, sulfate, volatile acidity, and total sulfur dioxide are significant. Next we are saving the results in a dataframe.

```
resultsDf = pd.DataFrame({'Method':['Decision Tree'],
'accuracy': acc_DT})
resultsDf = resultsDf[['Method', 'accuracy']]
resultsDf
```

Step 13: We will apply random forest now.

```
from sklearn.ensemble import RandomForestClassifier
rf_model = RandomForestClassifier(n_estimators = 50)
rf_model = rf_model.fit(X_train, y_train)
prediction_RF = rf_model.predict(X_test)
accuracy_RF = accuracy_score(y_test, prediction_RF)
tempResultsDf = pd.DataFrame({'Method':['Random Forest'],
'accuracy': [accuracy_RF]})
resultsDf = pd.concat([resultsDf, tempResultsDf])
resultsDf = resultsDf[['Method', 'accuracy']]
resultsDf
```

Step 14: We can compare the accuracies of both decision tree and random forest.

	Method	accuracy
0	Decision Tree	0.63125
0	Random Forest	0.76250

Step 15: We will now implement AdaBoost algorithm.

```
from sklearn.ensemble import AdaBoostClassifier
adaboost_classifier = AdaBoostClassifier( n_estimators= 150,
learning_rate=0.05, random_state=5)
adaboost_classifier = adaboost_classifier.fit(X_train, y_train)
prediction_adaboost =adaboost_classifier.predict(X_test)
accuracy_AB = accuracy_score(y_test, prediction_adaboost)
tempResultsDf = pd.DataFrame({'Method':['Adaboost'],
'accuracy': [accuracy_AB]})
resultsDf = pd.concat([resultsDf, tempResultsDf])
resultsDf = resultsDf[['Method', 'accuracy']]
resultsDf
```

	Method	accuracy
0	Decision Tree	0.63125
0	Random Forest	0.76250
0	Adaboost	0.63125

Step 16: We will implement bagging algorithm and compare the accuracies.

```
from sklearn.ensemble import BaggingClassifier

bagging_classifier = BaggingClassifier(n_estimators=55, max_
samples= .5, bootstrap=True, oob_score=True, random_state=5)
bagging_classifier = bagging_classifier.fit(X_train, y_train)
prediction_bagging =bagging_classifier.predict(X_test)
accuracy_bagging = accuracy_score(y_test, prediction_bagging)
tempResultsDf = pd.DataFrame({'Method':['Bagging'], 'accuracy':
[accuracy_bagging]})
resultsDf = pd.concat([resultsDf, tempResultsDf])
resultsDf = resultsDf[['Method', 'accuracy']]
resultsDf
```

	Method	accuracy
0	Decision Tree	0.631250
0	Random Forest	0.762500
0	Adaboost	0.631250
0	Bagging	0.734375

Step 17: We will now implement gradient boosting algorithm.

```
from sklearn.ensemble import GradientBoostingClassifier
gradientBoosting_classifier = GradientBoostingClassifier
(n_estimators = 60, learning_rate = 0.05, random_state=5)
```

```
gradientBoosting_classifier = gradientBoosting_classifier.
fit(X_train, y_train)
prediction_gradientBoosting =gradientBoosting_classifier.
predict(X_test)
accuracy_gradientBoosting = accuracy_score(y_test, prediction_
gradientBoosting)
tempResultsDf = pd.DataFrame({'Method':['Gradient Boost'],
'accuracy': [accuracy_gradientBoosting]})
resultsDf = pd.concat([resultsDf, tempResultsDf])
resultsDf = resultsDf[['Method', 'accuracy']]
resultsDf
```

	Method	accuracy
0	Decision Tree	0.631250
0	Random Forest	0.762500
0	Adaboost	0.631250
0	Bagging	0.734375
0	Gradient Boost	0.662500

We can deduce that random forest has given us the best accuracy as compared to the other algorithms. This solution can be extended to any supervised classification problem.

Gradient boosting is one of the most popular techniques. Its power is due to the focus it puts on errors and misclassifications. It is very useful in the field of information retrieval system where ML-based ranking is implemented. With its variants like extreme gradient boosting it can combat overfitting and missing variables and with CatBoost it overcomes the challenges with categorical variables. Along with bagging techniques, boosting is extending the predictive power of ML algorithms.

Note It is recommended to test random forest and gradient boosting while you are solving a real-world business problem since they offer higher flexibility and performance.

Ensemble methods are much more robust, accurate, and mature algorithms as compared to their counterparts. They enhance capabilities by combining the weak predictors and improve the overall performance. This is the reason they have outperformed other algorithms in many ML competitions. In the business world too, random forest and gradient boosting are frequently used to solve business problems.

We will now study another powerful algorithm called *Support Vector Machine* (*SVM*), which is often used for small but complex datasets having a large number of dimensions. It is a common challenge in industries like medical research where the dataset is generally small but has a very high number of dimensions. SVM serves the purpose very well and is discussed in the next section.

SVM

We have already studied classical ML algorithms like regression, decision tree, and so on in the previous chapters. They are quite competent to solve any sort of regression or classification problems for us and work on live datasets. But for "really" complex datasets, we require much higher capability. SVMs allow us those capabilities to process those multidimensional complex data sources. Complexity of the data source will be owing to the multiple dimensions we have and due to the different types of variables present in the data. Here, SVMs help in creating a robust solution.

SVM is a fantastic solution for complex datasets, particularly where we have a dearth of training examples. Apart from the uses on structured datasets and simpler business problems, it is used for categorization of text in text analytics problems, image classification, bioinformatics field, and handwriting recognition.

SVM can be used for both regression and classification problems. The basis of SVM is on support vectors which are nothing but the representation of observations in a vector space.

The way to visualize is in Figure 4-3. Imagine we have a dataset with "n" attributes. These n features can hence be represented in an n-dimensional space, where values of each attribute refer to the coordinates. In Figure 4-3, we are representing only 2-dimensional space. A similar representation can be made for an n-dimensional space.

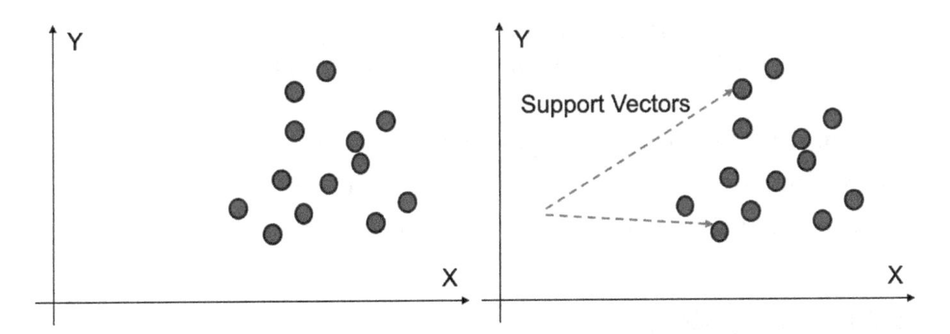

***Figure 4-3.** Support vectors are the representations of data points in a vector-space diagram*

SVMs work on these representations or support vectors and model a supervised learning algorithm. In Figure 4-4, we have two classes which need to be differentiated. SVM fixes this problem by creating hyperplane, which is most suitable for that decision.

As shown in Figure 4-4, the distance between the nearest data point and the hyperplane is called the *margin*. In the problem, SVM finds a linear plane with maximum margin to be able to distinguish between the classes clearly. Unlike the linear classifiers where we want to minimize the sum of squares of errors, in SVM the objective is to find that linear plane which distinguishes two or more classes and separates them with maximum margin. SVM implementation can also be referred to as *maximum-margin hyperplane.*

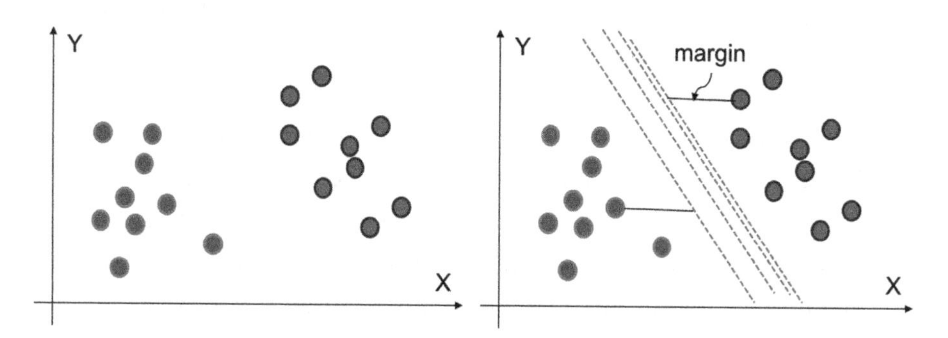

Figure 4-4. *A hyperplane can be used to distinguish between two classes. Margin is used to select the best hyperplane*

We have understood the purpose of the SVM. It is imperative we visualize them in a vector-space diagram to understand better. We are visualizing SVM in a 2-dimensional space in the next section.

SVM in 2-D Space

In a 2-dimensional space, the separating hyperplane is a straight line. And the classification is achieved by a perceptron.

A perceptron is an algorithm used to make binary classification. Simply put, it is trained on data with two classes and then it outputs a line that separates two classes clearly. In Figure 4-5, we can try to achieve a hyperplane or a line which segregates the two classes.

Now there can be multiple hyperplanes which can serve the purpose of generating the correct classifications. As shown in Figure 4-5, the first line separates the two classes clearly, but it is very close to the two classes or red and blue dots in this case. Though it is good for classification, this model is susceptible to higher *variance* if we deploy the model into production on a new, unseen dataset, due to which few of the observations will be classified wrong. The second line does not suffer from such an issue. The second line is at the maximum distance from both the classes simultaneously and hence will be selected.

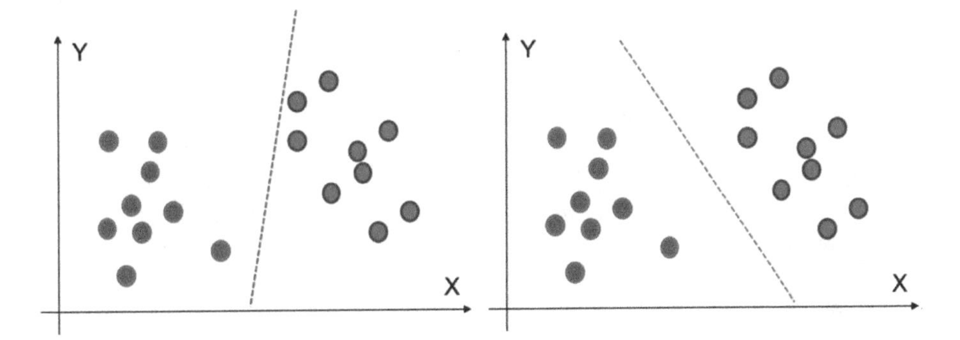

Figure 4-5. *The red line, though able to classify between two classes, suffers from high variance. The black classifier on the right is better than the red one*

So, we have decided that the second line is better than the first line. Let us say that the equation of the line is ax + by = c. Hence, the equation for the classification plane can be ax + by ≥ c for the red dots and ax + by < c for the blue ones.

But there can be a lot of options for a, b, or c, which brings us to the next question: how to choose the best plane. As shown in Figure 4-6, there can be a number of options available for the hyperplanes.

In Figure 4-6, for the figure on top left, the red separator is doing a better job in classifying the two classes than the black solid line. In the second figure, we can see that the red separator has a maximum margin as compared to the black one, hence it is chosen.

The third case shows the presence of a few outliers in the dataset. Still the SVM algorithm will be able to create a classification hyperplane with maximum margin. SVM works quite well even in the presence of outliers.

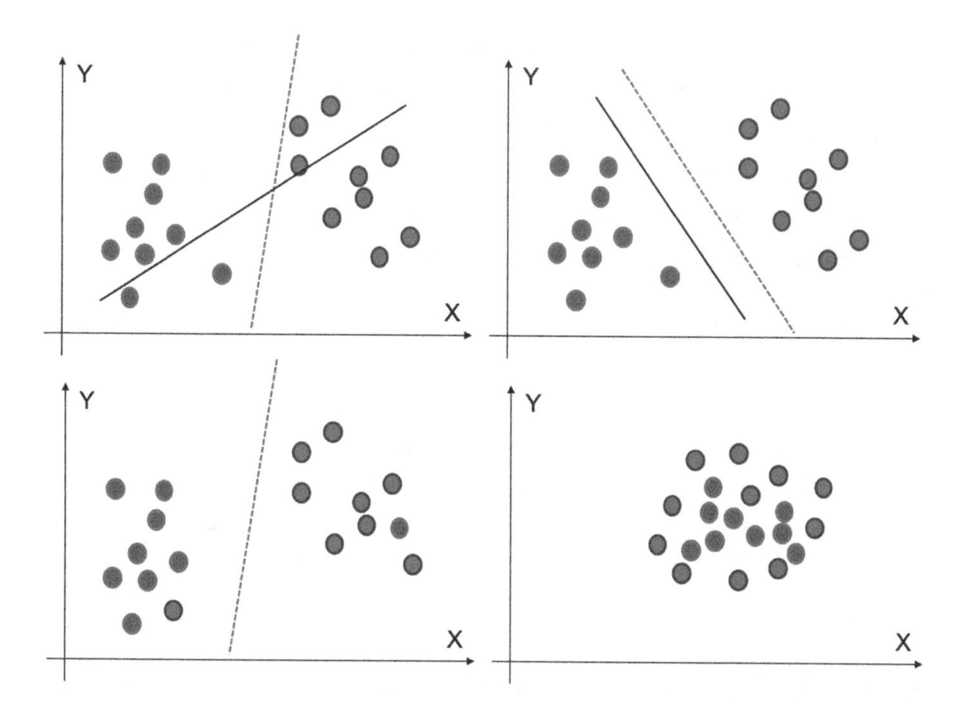

Figure 4-6. *(i) At top left, the red classifier is better than the black one. (ii) In the second one, red is better as it has maximum margin. (iii) The third one has outliers but still SVM will be able to handle it. (iv) This is a special case where linear classifier will not be able to distinguish between the two classes*

So far, we have discussed and visualized the implementations in a 2-D space, but if were to transform the mathematical vector space from a 2-dimensional one to higher dimensions, then we need to perform such mathematical operations. In Figure 4-6, the fourth figure is such a special case. In the case shown, it is not possible to have a linear hyperplane, and in such a case we will have a nonlinear hyperplane to make the classifications for us, which is possible using *kernel SVM (KSVM)*, which we are discussing next.

KSVM

If we transform 2-dimensional space into high-dimensional space, the solution becomes more robust and the respective probability to separate the data points increases. For example, x_1 and x_2 have to be converted into higher degrees of polynomials x_1^2, x_2^2, x_3^2, x_4^2, x_5^2, and so on. This is achieved by kernel or KSVM. It takes the data points into a higher-order mathematical space. In this high-dimensional space, they become separable linearly. Then we are able to draw a plan through these data points. If we represent the previous example using KSVM, the corresponding representation will be as shown in Figure 4-7.

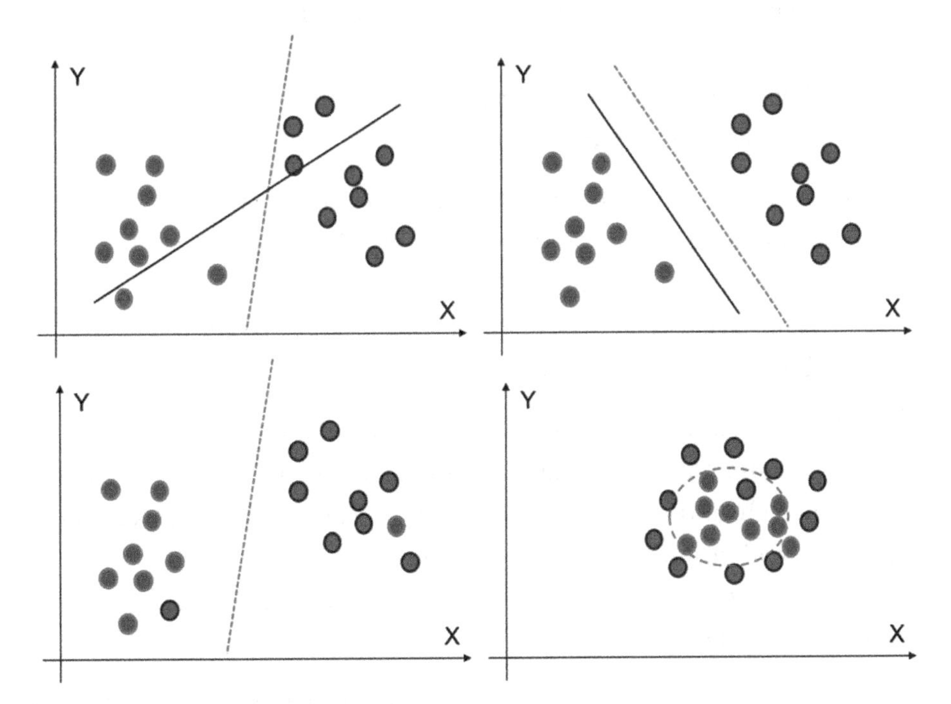

Figure 4-7. *The fourth diagram is implementing a nonlinear classifier to distinguish between two classes*

KSVM has created a nonlinear classifier to perform the classification between the two classes.

Here are some of the parameters of SVM:

1. **Kernel**: Kernel is used when we have the data which can become separable if expressed in higher dimensions. The various kernels available in sklearn are rbf, poly, sigmoid, linear, precomputed, and so on. If we use a "linear" kernel, it will use a linear hyperplane or a line in the case of 2-dimensional data. 'rbf', 'poly' are used for nonlinear hyperplanes.

2. **C**: C is used to represent the misclassification error or the *cost* parameter. If the value of C is low, the penalty of misclassification observations is low and hence the accuracy will be high. It is used to control the tradeoff between accurate classification of training data and having a smooth decision boundary.

3. **Gamma**: Gamma is used to define the radius of influence of the observations in a classification. It is primarily used for nonlinear hyperplanes. A higher gamma can lead to better accuracy, but results can be biased, and vice versa.

We have to iterate with various values of such parameters and reach the best solution. With a high value of gamma, the variance will be low and bias will be high, and vice versa. And when the value of C is large, variance will be high and bias will be low, and vice versa.

There are both advantages and some challenges with using SVM.

Advantages of SVM solution:

1. It is a very effective solution for complex datasets, where the number of dimensions is large.

2. It is the preferred choice when we have more dimensions and less training dataset.

3. The margin of separation by SVM is quite clear and provides a good, accurate, and robust solution.
4. SVM is easy to implement and is quite a memory-efficient solution to implement.

Challenges with SVM:

1. It takes time to converge with large sample size and hence may not be preferred for bigger datasets.
2. The algorithm is sensitive to messy data. If the target classes are not clearly demarcated and different, the algorithm tends to perform not so well.
3. SVM does not provide direct probabilities for the predictions. Instead they have to be calculated separately.

Despite a few challenges, SVM has repeatedly proven its worth. It offers a robust solution when we have a multidimensional smaller data set to analyze. We are now going to solve a case study in Python using SVM now.

Case Study Using SVM

We are solving a cancer detection case study. The dataset is available at the Github link shared at the start of the chapter.

Step 1: Import the necessary libraries first

```
import pandas as pd
import numpy as np
import seaborn as sns
import matplotlib.pyplot as plt
%matplotlib inline
```

Step 2: Import the dataset now

```
cancer_data = pd.read_csv('bc2.csv')
cancer_dataset = pd.DataFrame(cancer_data)
cancer_dataset.columns
```

```
cancer_data = pd.read_csv('bc2.csv')
cancer_dataset = pd.DataFrame(cancer_data)
cancer_dataset.columns

Index(['ID', 'ClumpThickness', 'Cell Size', 'Cell Shape', 'Marginal Adhesion',
       'Single Epithelial Cell Size', 'Bare Nuclei', 'Normal Nucleoli',
       'Bland Chromatin', 'Mitoses', 'Class'],
      dtype='object')
```

Step 3: Have a look at the columns

```
cancer_dataset.describe()
```

```
cancer_dataset.describe()
```

	ID	ClumpThickness	Cell Size	Cell Shape	Marginal Adhesion	Single Epithelial Cell Size	Normal Nucleoli	Bland Chromatin	Mitoses	Class
count	6.990000e+02	699.000000	699.000000	699.000000	699.000000	699.000000	699.000000	699.000000	699.000000	699.000000
mean	1.071704e+06	4.417740	3.134478	3.207439	2.806867	3.216023	3.437768	2.866953	1.589413	2.689557
std	6.170957e+05	2.815741	3.051459	2.971913	2.855379	2.214300	2.438364	3.053634	1.715078	0.951273
min	6.163400e+04	1.000000	1.000000	1.000000	1.000000	1.000000	1.000000	1.000000	1.000000	2.000000
25%	8.706885e+05	2.000000	1.000000	1.000000	1.000000	2.000000	2.000000	1.000000	1.000000	2.000000
50%	1.171710e+06	4.000000	1.000000	1.000000	1.000000	2.000000	3.000000	1.000000	1.000000	2.000000
75%	1.238298e+06	6.000000	5.000000	5.000000	4.000000	4.000000	5.000000	4.000000	1.000000	4.000000
max	1.345435e+07	10.000000	10.000000	10.000000	10.000000	10.000000	10.000000	10.000000	10.000000	4.000000

Step 4: Treat the missing values in the next step. We are filling them with median here.

```
cancer_dataset = cancer_dataset.replace('?', np.nan)
cancer_dataset = cancer_dataset.apply(lambda x: x.fillna(x.
median()),axis=0)
```

Step 5: Now we are converting the 'Bare Nuclei' column string type to float

```
cancer_dataset['Bare Nuclei'] = cancer_dataset['Bare Nuclei'].
astype('float64')
```

Step 6: Check if there are any NULL values present in the dataset

```
cancer_dataset.isnull().sum()
```

```
cancer_dataset.isnull().sum()

ID                             0
ClumpThickness                 0
Cell Size                      0
Cell Shape                     0
Marginal Adhesion              0
Single Epithelial Cell Size    0
Bare Nuclei                    0
Normal Nucleoli                0
Bland Chromatin                0
Mitoses                        0
Class                          0
dtype: int64
```

Step 7: Divide the data into train and test now.

```
from sklearn.model_selection import train_test_split

# To calculate the accuracy score of the model
from sklearn.metrics import accuracy_score, confusion_matrix

target_variable = cancer_dataset["Class"]
features = cancer_dataset.drop(["ID","Class"], axis=1)
X_train, X_test, y_train, y_test = train_test_
split(features,target_variable, test_size = 0.25,
random_state = 5)
```

Step 8: Train the model with linear kernel

```
from sklearn.svm import SVC
svc_model = SVC(C= .1, kernel='linear', gamma= 1)
svc_model.fit(X_train, y_train)
svc_prediction = svc_model .predict(X_test)
```

Step 9: Check the accuracy

```
print(svc_model.score(X_train, y_train))
print(svc_model.score(X_test, y_test))
```

```
# check the accuracy on the training set
print(svc_model.score(X_train, y_train))
print(svc_model.score(X_test, y_test))

0.9751908396946565
0.9485714285714286
```

Step 10: Print the confusion matrix

```
print("Confusion Matrix:\n",confusion_matrix(svc_prediction,
y_test))
```

```
print("Confusion Matrix:\n",confusion_matrix(svc_prediction,y_test))

Confusion Matrix:
 [[108   4]
 [  5  58]]
```

Step 11: In the next steps, we will change the kernel and get different accuracies

```
svc_model = SVC(kernel='rbf')
svc_model.fit(X_train, y_train)
```

```
print(svc_model.score(X_train, y_train))
print(svc_model.score(X_test, y_test))
```

```
print(svc_model.score(X_train, y_train))
print(svc_model.score(X_test, y_test))

0.9751908396946565
0.96
```

```
svc_model  = SVC(kernel='poly')
svc_model.fit(X_train, y_train)
svc_prediction = svc_model.predict(X_test)
print(svc_model.score(X_train, y_train))
print(svc_model.score(X_test, y_test))
```

```
print(svc_model.score(X_train, y_train))
print(svc_model.score(X_test, y_test))

0.982824427480916
0.9371428571428572
```

```
svc_model = SVC(kernel='sigmoid')
svc_model.fit(X_train, y_train)
svc_prediction = svc_model.predict(X_test)
print(svc_model.score(X_train, y_train))
print(svc_model.score(X_test, y_test))
```

```
print(svc_model.score(X_train, y_train))
print(svc_model.score(X_test, y_test))

0.4541984732824427
0.42857142857142855
```

We can compare the respective accuracies for all the kernels and choose the best one. Ideally, accuracy should not be the only parameter; we should also compare recall and precision using confusion matrix.

In the preceding example, we can create a visualization using seaborn library. It is an additional step and can be done if needed.

```
sns.pairplot(cancer_dataset, diag_kind = "kde", hue = "Class")
```

In the preceding example, we created a Python solution using SVM. When we change the kernel, the accuracy changes a lot. The SVM algorithm should be compared along with other ML models and then the best algorithm should be chosen.

Note Ideally we test any problem with three or four algorithms and compare the precision, recall, accuracy and then decide which algorithm is best for us. These steps are discussed again in Chapter 5.

With this we have studied the SVM in detail. An easy-to-implement solution, SVM is one of the advanced supervised learning algorithms which is heavily recommended.

So far, we have studied and created solutions for structured data. We started with regression, decision trees, and so on in previous chapters. We examined the concepts and created a solution in Python. In this chapter, we continued with boosting algorithms and SVMs. Now we will start a much more advanced topic—supervised learning algorithms for unstructured data, which are text and images, in the next section. We will study the nuts and bolts, preprocessing steps, challenges faced, and use cases. And like always, we will create Python solutions to complement the knowledge.

Supervised Algorithms for Unstructured Data

We now have access to cameras, phone, processors, recorders, data management platforms, cloud-based infrastructure, and so on. And hence, our capabilities to record data, manage it, store it, transform it, and analyze it have also improved tremendously. We are not only able to capture complex datasets but also store them and process them. With the advent of neural network–powered deep learning, the processing has improved drastically. Deep learning is a revolution in itself. Neural networks are fueling the limitless capabilities being developed across domains and business. With superior processing powers, and more powerful machines like multicore GPU and TPU, sophisticated deep neural networks are able to process more information much faster, which is true for both structured and unstructured datasets. In this section, we are going to work on unstructured datasets and study supervised learning algorithms for unstructured datasets.

Recall in Chapter 1, we discussed structured and unstructured datasets as shown in Figure 4-8. Text, images, audio, video, and so on fall into the unstructured datasets.

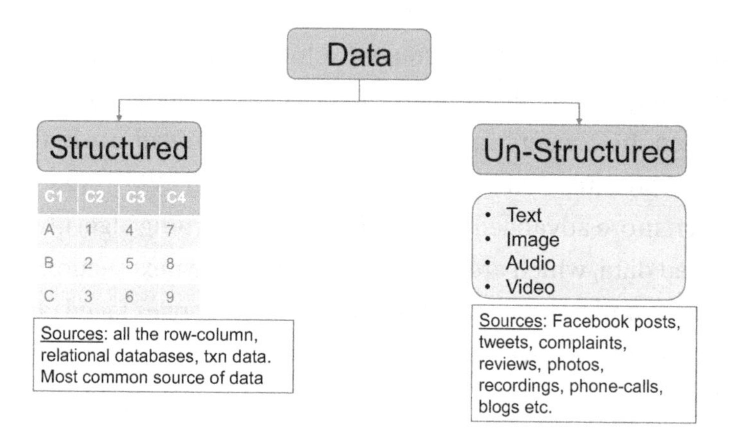

Figure 4-8. *Data can be classified between structured and unstructured datasets*

We will start studying the text data. We will examine all the concepts of cleaning the text data, preprocessing it, creating supervised learning solutions in Python using it, and what are the best practices to work with it. Let's kick off now!

Text Data

Language is a gift to humanity. It is the most common medium to express ourselves. Language is involved in most interactions we have like speaking, messaging, writing, listening. This text data is everywhere. We generate it every day in the form of news, Facebook comments and posts, customer reviews and complaints, tweets, blogs, articles, literature, and so on. The datasets generated represent a wide range of emotions and expressions which are generally not captured in surveys and ratings. We do witness that in the form of online product reviews given by customers. The number of ratings given can be 5 out of 5, but the actual review text might give a different impression. Thus, it becomes even more crucial for businesses worldwide to pay attention to the text data.

Text data is much more expressive and direct. This data is to be analyzed as it holds the key to a lot of understanding we can generate about our customers, processes, products and services, our culture, our world, and our thoughts. Moreover, with the advent of Alexa, Google Assistant, Apple Siri, and Cortana the voice command is acting as an interface between humans and machines and generating more datasets for us. Massive and expressive, right!

Similar to the complexity, text data is a rich source of information and actions. Text data can be used for a plethora of solutions, which we discuss next.

Use Cases of Text Data

Text data is very useful. It expresses what we really feel in words. It is a powerful source to gauge the thoughts which often are not captured in surveys and questionnaires. It is directly sourced data and hence is less biased, though it can be a really noisy dataset to deal with.

Text data is quite rich and can be used for multiple use cases like the following:

1. **News categorization or document categorization**: We can have incoming news or a document. We want to categorize whether a news item belongs to sports, politics, science, business, or any other category. Incoming news will be classified based on the content of the news, which is the actual text.

News about business will be different from a news article on sports as shown in Figure 4-9. Similarly, we might want to categorize some medical documents into their respective categories based on the domain of study. For such purposes, supervised learning classification algorithms can be used to solve the problems.

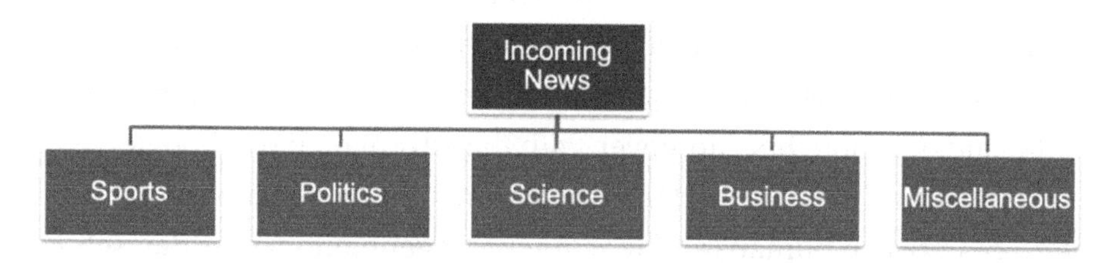

Figure 4-9. *There can be multiple categories of incoming news as sports, politics, science, business, and so on*

2. **Sentiment analysis**: Sentiment analysis is gauging what is the positiveness or negativity in the text data. There can be two such use cases:

    a. We receive reviews from our customers about the products and services. These reviews have to be analyzed. Let's consider a case. An electric company receives complaints from its customers, reviews about the supply, and comments about the overall experience. The streams can be onboarding experience, ease of registration, payment process, supply reviews, power reviews, and so on. We want to determine the general context of the review—whether it is positive, negative, or neutral. Based on these comments, an improvement can be made on the product features or service levels.

   b. We might also want to assign a review to a specific department. For example, in the preceding case an incoming review will have to be shared with the relevant department. Using Natural Language Processing (NLP), this task can be done, the review can be shared with the finance department or operations team, and the respective team can follow up with the review and take the next course of action.

3. **Language translation**: Using NLP and deep learning, we are able to translate between languages (e.g., between English and French). A deep neural network requires training on the vocabulary and grammar of both the languages and multiple other training data points.

4. **Spam filtering**: Email spam filter can be composed using NLP and supervised ML. We can train an algorithm which can analyze incoming mail parameters and give a prediction if that email belongs to a spam folder or not. Going even one step further, based on the various parameters like sender email-id, subject line, body of the mail, attachments, time of mail, and so on, we can even determine if that is a promotional email or spam or an important one. Supervised learning algorithms help us in making that decision and automating the entire process.

5. **Text summarization** of the entire book or article can be done using NLP and deep learning. In this case too, we will be using deep learning and NLP to generate summaries of entire documents or articles. This helps in creating us an abridged version of the text.

6. **Part-of-speech (POS) tagging**: POS tagging refers to the identification of words as nouns, pronouns, adjectives, adverbs, conjunctions, and so on. It is the process of marking the words in the text corpus corresponding to a particular POS, based on its use, definition, and context in the sentence and larger body, as shown in Figure 4-10.

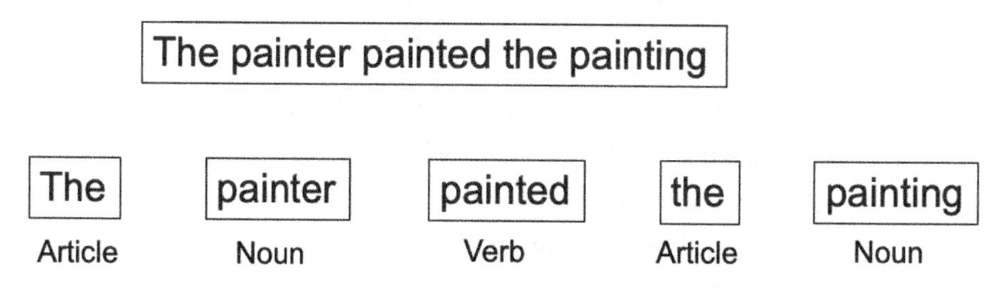

Figure 4-10. *POS tagging for words into their respective categories*

Text data can be analyzed using both supervised and unsupervised problems. We are focusing on supervised learning in this book. We use NLP to solve the problems. And deep learning further improves the capabilities we have. These are the tools which empower us to deal with such complex datasets.

But text data is difficult to analyze. It has to be still represented in the form of numbers and integers; only then can it be analyzed. Our computers and processors understand numbers and the algorithms also expect numbers only. We will now discuss the most common challenges we face with text data in the next section.

Challenges with Text Data

Text is perhaps the most difficult data to be analyzed and worked with. The number of permutations to express the same question or thought are many. For example, “what is your age” and “how old are you” mean one and the same thing. We have to resolve these challenges and come up with a dataset which is robust, complete, and representative while at the same time not losing the original context.

The most common challenges we face are as follows:

1. Language is unbounded. It changes every day and every moment new words are added to the dictionaries.

2. Languages are many: Hindi, English, French, Spanish, German, Italian, and so on. Each language follows its own rules and grammar, which are unique in usage and pattern. Some are written left to right; some might be right to left or maybe even vertically! A thought which is expressed in twelve words in one language might be expressed in only five words in another.

3. A word can change its meaning in a different context. For example, "I want to read this book" and "Please book the hotel for me." A word can be an adjective and can be a noun too depending on the context.

4. A language can have many synonyms for the same word; for example, "good" can be replaced by "positive," "wonderful," "superb," and "exceptional" in different scenarios. Similarly, words like "study," "studying," and "studies" are related to the same root word, "study."

5. Words can completely even change their meaning with usage. For example, "apple" is a fruit, while "Apple" is a company producing Macintosh. "Tom" can be a name but when used as "Tom Software Consulting," its usage is completely changed.

6. Tasks which are very easy for humans might be very difficult for machines. We do have memory, while machines tend to forget. For example, "John is from London and he moved to Australia and is working with Allan over there. He missed his family back there." Humans can easily recall and understand that "he" in the second sentence is John and not Allan.

These are not the only challenges, and the preceding list is not exhaustive. Managing this massive dataset, storing it, cleaning it, and refreshing it is a Herculean task in itself. But using sophisticated processes and solutions, we are able to resolve most of them, if not all. Some of these techniques we discuss in the next section on preprocessing the text data and extracting the features from the text data.

Like any other ML project, text analytics follow the principles of ML, albeit the process is slightly different. We will discuss the text analytics process now in the next section.

Text Analytics Modeling Process

Text analytics is complex owing to the complexity of data we are dealing with and the data preprocessing required. At a high level the various process heads remain the same, but still a lot of the subprocesses are customized for text. They are also dependent on the business problem we want to solve. The typical text analytics process is shown in Figure 4-11.

Text analytics process, similar to any other project, starts with the definition of a *business problem*. The business problem can be the use cases which we have discussed in the previous section, which can be sentiment analysis or text summarization.

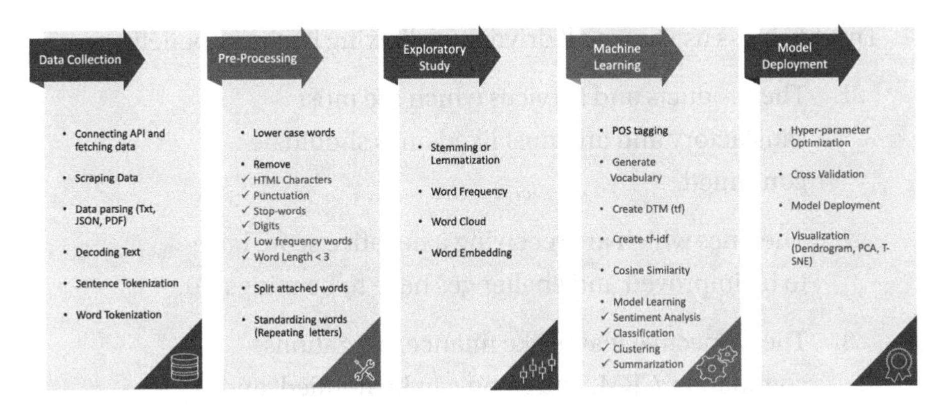

***Figure 4-11.** End-to-end process in a text analytics project from data collection to the deployment*

Let's consider the same business problem we discussed in the last section: sentiment analysis for an electric company. The business problem can be that we receive customer complaints, reviews about our products, and services through many mediums like call centers, emails, phone calls, messages, and so on. At the same time, many customers post on Facebook or other social media platforms. This generates a lot of text data for us. We want to analyze this text data and generate findings and insights about

1. Our customer's satisfaction regarding the products and services.
2. The major pain points and dissatisfactions, what drives the engagement, which services are complex and time-consuming, and which are the most liked services.
3. The products and services which are most popular, which are least popular, and any popularity patterns.
4. How best to represent these findings by means of a dashboard. This dashboard will be refreshed at a regular cycle like monthly or quarterly refresh.

This business use case will drive the following business benefits:

1. The products and services which are most satisfactory and are most liked ones should be continued.
2. The ones which are receiving a negative score have to be improved and challenges have to be mitigated.
3. The respective teams like finance, operations, complaints, CRM, and so on can be notified, and they can work individually to improve the customer experience.
4. The precise reasons for liking or disliking the services will be useful for the relevant teams to work in the correct direction.
5. Overall, it will provide a benchmark to measure the Net Promoter Score (NPS) for the customer base. The business can strive to enhance the overall customer experience.

A concise, precise, measurable, and achievable business problem is the key to success. Once the business problem is frozen, we will work on securing the dataset, which we will discuss next.

Text Data Extraction and Management

As discussed in the last section, customer text data can be generated through a number of sources. The entire dataset in text analytics is referred to as a *corpus*. Formally put, a corpus represents a large collection of text data (generally labeled but can be unlabeled too), which is used for statistical analysis and hypothesis testing.

In Figure 4-12, we are depicting a process of receiving the text data from multiple sources like call center calls, complaints, reviews, blogs, tweets, and so on. These data points are first moved to a staging area.

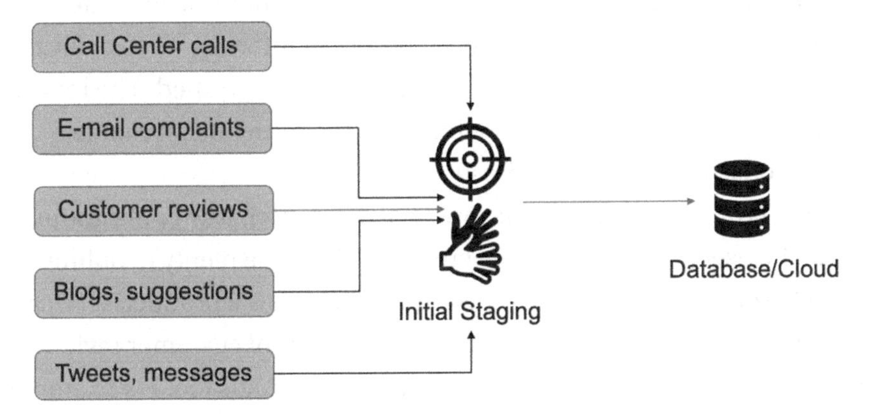

***Figure 4-12.** Data management process of text data, starting from data collection to the final storage*

To be noted is that these various data sources might have different types of data points like .csv, .xls, .txt, logs, files or databases, json or .pdf, and so on. We might even fetch data from APIs. During the staging area load, all of these data points have to be merged and cleaned. It will involve creation of databases, tables, views, and so on. In the case study shared previously, the database table can look like the following structure in Table 4-2.

***Table 4-2.** Customer Reviews Having Customer Details Like ID, Date, Product, City, and Actual Review Text*

CustID	Product	Date	City	Source	ReviewText
1001	ABC	01-Jan-20	London	Tweet	....
1002	XYZ	02-Jan-20	London	FB	....
1003	ABC	03-Jan-20	London	Email	....
1004	ABC	04-Jan-20	London	Call Centre	....
1005	XYZ	05-Jan-20	London	Other	....

Here, we have the unique customer ID, product purchased, date of the review, city, source, and actual review text of the customer. This table can have many more data points then we have shown. And there can be other tables having customer details like was the complaint resolved, time to resolve, and so on, which can serve as additional information to analyze.

All such data points have to be maintained and refreshed. The refresh cycle can be determined as per the business requirements: it can be a monthly, quarterly, or yearly refresh.

There is one more important source of data which is very insightful and can be used for wider strategy creation. There are plenty of online channels and platforms where a customer can review a product/service. Online marketplaces like Amazon also have details of customer reviews. These platforms have reviews of the competitive brands too. For example, Nike might be interested in Puma's and Reebok's customer reviews. These online reviews have to be scraped and maintained. Again, these reviews might be in a different format and will have to be cleaned.

During this data maintenance phase, we clean a lot of text data like junk characters like *&^# which are present in the data. They might occur because of formatting errors while loading the data or the data itself might have junk characters. The text data is cleaned to the maximum possible extent, and further cleaning can take place in the next step of data preprocessing.

Text data is really tough data to deal with. There are a lot of complexities and data is generally messy. We have discussed a few of the challenges in the last section. We will be examining a few again and examining solutions to tackle them in the next section. We are starting with extracting features from the text data, representing them in a vector-space diagram and creating ML models using these features.

Preprocessing of Text Data

Text data, like any other data source, can be messy and noisy. We clean some of it in the data discovery phase and a lot of it in the preprocessing phase. At the same time, we have to extract the features from our dataset. This cleaning process is a standard one and can be implemented on most of the text datasets.

There are multiple processes in which we complete these steps. We will start with cleaning the raw text first.

Data Cleaning

There is no second thought about the importance of data quality. The cleaner the text data is, the better the analysis will be. At the same time, reducing the size of the text data will result in a lower-dimensional data. And hence, the processing during the ML phase and training the algorithms become less complex and time-consuming.

Text data is to be cleaned as it contains a lot of junk characters, irrelevant words, noise and punctuation, URLs, and so on. The primary ways of cleaning the text data are

1. **Stop-word removal**: Stop words are the most common words in a vocabulary which carry less importance than the keywords. For example, "is," "an," "the," "a," "be," "has," "had," "it," and so on. It reduces the dimensions of the data and hence complexity is reduced. But due caution is required while we remove stop words. For example, if we ask the question "Is it raining?" then the answer "It is" is a complete answer in itself.

Note When we are working with problems where contextual information is important like machine translation, we should avoid removing stop words.

2. **Library-based cleaning**: This involves cleaning of data based on a predefined library. We can create a repository of words which we do not want in our text and can iteratively remove from the text data. This approach is preferred if we do not want to use a stop-word approach but want to follow a customized one.

3. **Junk characters**: We can remove URL, hashtags, numbers, punctuations, social media mentions, special characters, and so on from the text. We have to be careful as some words which are not important for one domain might be quite useful for a different domain.

Note Due precaution is required when data is cleaned. We have to always keep the business context in mind while we remove words or reduce the size.

4. **Lexicon normalization**: Depending on the context and usage, the same word might get represented in different manners. During lexicon normalization we clean such ambiguities. The basic idea is to reduce the word to its root form. Hence, words which are derived from each other can be mapped to the central word provided they have the same core meaning.

For example, study might get represented as study, studies, studied, studying, and so on as shown in Figure 4-13. The root word "study" remains the same, but its representations differ.

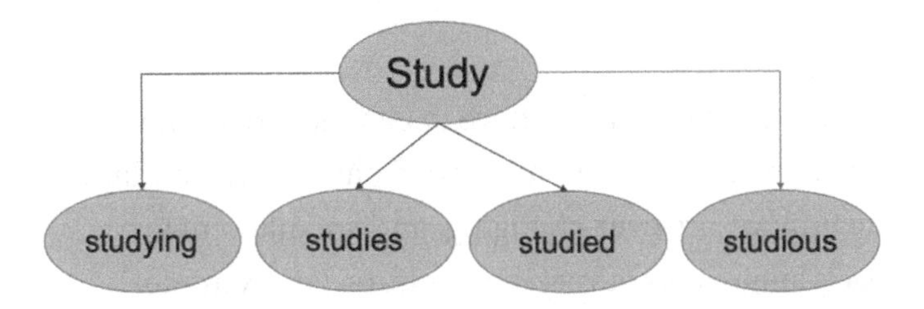

Figure 4-13. *The root word is "study," but there are many forms of it like "studying" and "studies"*

There are two ways to deal with this, namely, stemming and lemmatization:

a. *Stemming* is a very basic rule-based approach of removing "es," "ing," "ly," "ed," and so on from the end of the word. For example, "studies" will become "studi" and "studying" will become "study." As visible being a rule-based approach, the output spellings might not always be accurate.

b. In contrast to stemming, *lemmatization* is an organized approach which reduces words to their dictionary form. A *lemma* of a word is its dictionary or canonical form. For example, "studies," "studied," and "studying" all have the same root word, "study."

5. **Standardization**: With the advent of modern communication devices and social media, our modes of communications have changed. Along with it, our language has also changed. We have new limitations and rules, like a tweet can be of 280 characters only.

   Hence, the dictionaries have to change too. We have newer references which are not a part of any standard dictionary, are ever-changing, and are different for each language, country, and culture. For example, "u" refers to "you," "luv" is "love," and so on.

   We have to clean such text too. In such a case, we create a dictionary of such words and replace them with the correct full form.

These are only some of the methods to clean the text data. These techniques should resolve most of the issues. Still, we will not get completely clean data. Business acumen is required to further make sense to it.

Once the data is cleaned, we have to start representation of data so that it can be processed by ML algorithms—our next topic.

Extracting Features from Text Data

Text data, like any other data source, can be messy and noisy. We clean some of it in the data discovery phase and in the preprocessing phase. Now the data is clean and ready to be used. The next step is to represent this data in a format which can be understood by our algorithms.

In the simplest understanding, we can simply perform *one-hot encoding* on our words and represent them in a matrix. The words can be first converted to lowercase and then sorted in an alphabetical order. And then a numeric label can be assigned. And finally, words are converted to binary vectors. We will explain it using an example.

For example, the text is "He is going outside." We will use the following steps:

1. We will convert the words to lowercase, resulting in - he, is, going, outside.

2. Next, arrange the words in alphabetical order, which gives the output as - going, he, is, outside.

3. We can now assign values to each word as going:0, he:1, is:2, outside:3.

4. Finally, they are transformed to binary vectors as

   [[0. 1. 0. 0.] #he

   [0. 0. 1. 0.] #is

   [1. 0. 0. 0.] #going

   [0. 0. 0. 1.]] #outside

Though this approach is quite intuitive and simple to comprehend, it is pragmatically not possible due to the massive size of the corpus and the vocabulary. Moreover, handling such data size with so many dimensions will be computationally very expensive. The resulting matrix thus created will be very sparse too. Hence, we look at other means and ways to represent our text data.

There are better alternatives available to one-hot encoding. These techniques focus on the frequency of the word or the context in which the word is being used. This scientific method of text representation is much more accurate, robust, and explanatory. It generates better results too.

There are multiple such techniques like tf-idf, bag-of-words (BOW) approach, and so on. We discuss a few of these techniques in the next sections. But we will examine the important concept of tokenization first!

Tokenization

Text data has to be analyzed, and hence we can represent the words as *tokens*. Tokenization is breaking a text or a set of text into individual tokens. It is the building block of NLP. Tokens are usually individual words, but this is not necessary. We can tokenize a word or subwords or characters in the word. We can represent word tokenization as shown in Figure 4-14.

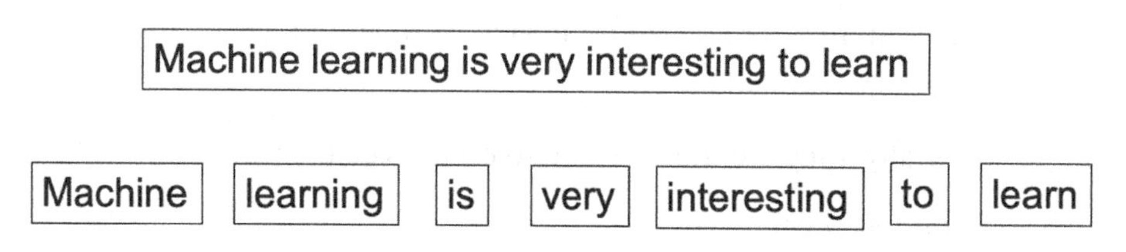

***Figure 4-14.** Tokenization of a sentence results in individual tokens for all the words*

In the case of subwords, the same sentence can have subword tokens as interest-ing. For tokenization at a character level, it can be i-n-t-e-r-e-s-t-i-n-g. In fact, in the one-hot encoding approach discussed in the last section as a first step, tokenization was done on the words.

There are multiple methods of tokenizing based on the regular expressions to match either tokens or separators between tokens. *Regexp* tokenization uses the given pattern arguments to match the tokens or separators between the tokens. *Whitespace* tokenization treats any sequence of whitespace characters as a separator. Then we have *blankline* which uses a sequence of blank lines as a separator. And *wordpunct* tokenizes by matching sequence of alphabetic characters and sequence of non-alphabetic and non-whitespace characters.

Tokenization hence allows us to assign unique identifiers or tokens to each of the words. These tokens are further useful in the next stage of the analysis.

Now, we will explore more methods to represent text data. The first such method is the "bag of words."

Bag-of-Words Model

In the bag-of-words approach, or BOW, text is tokenized for each observation it finds and then the respective frequency of each token is calculated. This is done disregarding grammar or word order; the primary goal is to maintain simplicity. Hence, we will represent each text (sentence or a document) as a *bag of its own words.*

Figure 4-15 shows that in the first example, each word has occurred only once and hence the frequency is 1. In the second sentence, the frequency of "is" and "drinking" is 2. This is called the bag of words for each token.

Machine learning is very interesting to learn

Word	Frequency
Machine	1
learning	1
is	1
very	1
interesting	1
to	1
learn	1

Tom is drinking milk while Jack is drinking coffee

Word	Frequency
Tom	1
is	2
drinking	2
milk	1
while	1
Jack	1
coffee	1

Figure 4-15. *Bag-of-words approach showing that words with higher frequency are given higher values*

In the BOW approach for the entire document, we define the vocabulary of the corpus as all the unique words present in the corpus. We can also set a threshold, that is, the upper and lower limit for the frequency. Then each sentence or document is defined by a vector of the same dimension as the base vocabulary containing the frequency of each word of the vocabulary in the sentence.

For example, if we imagine that the last two sentences—"Machine learning is very interesting to learn" and "Tom is an eating apple while Jack is drinking coffee"—as the only two sentences present in the entire vocabulary, then we will represent the first sentence as shown in Figure 4-16. We should note that drinking and milk are given 0 in this vector for "Machine learning is very interesting to learn."

Machine learning is very interesting to learn

Word	Frequency
Machine	1
learning	1
drinking	0
milk	0
is	1
very	1
interesting	1
to	1
learn	1
.....	.....

Figure 4-16. *Bag-of-words representation of a sentence based on the entire vocabulary*

The BOW approach has not considered the order of the words or the context. It focuses only on the frequency of the word. Hence, it is a very fast approach to represent the data. Since it is frequency based, it is commonly used for document classifications. At the same time, owing to pure frequency-based methods the model accuracy can take a hit. And that is why we have other advanced methods which consider more parameters then frequency alone. One of such methods is tf-idf or term frequency and inverse-document frequency, which we are studying next.

Term-Frequency and Inverse-Document Frequency

In the bag-of-words approach, we gave importance to the frequency of a word only. In term-frequency and inverse-document-frequency (tf-idf), we consider the relative importance of the word. tf-idf is made up of tf (term frequency) and idf (inverse-document frequency).

Term frequency (tf) is the count of a term in the entire document. For example, the count of the word "x" in the document "d."

Inverse-document frequency (idf) is the log of the ratio of total documents (N) in the entire corpus and number of documents (df) which contain the word "x."

So, the tf-idf formula will give us the relative importance of a word in the entire corpus. It is a multiplication of tf and idf and is given by

$$w_{i,j} = tf_{i,j} \times \log(N/df_i) \qquad \text{(Equation 4-1)}$$

where N is the total number of documents in the corpus

$tf_{i,j}$ is the frequency of the word in the document

df_i is the number of documents in the corpus which contain that word.

Let's understand this with an example.

Consider we have a collection of 1 million medical documents. In these documents, we want to calculate tf-idf value for the words "medicine" and "penicillin."

Let's assume that there is a document of 100 words having "medicine" five times and "penicillin" only twice. So tf for "medicine" is 5/100 = 0.05 and for "penicillin" is 2/100 = 0.02.

Now, we assume that "medicine" appears in 100,000 documents out of 1 million documents, while "penicillin" appears only in 10. So, idf for "medicine" is log (1,000,000/100,000) = log (10) = 1. For "penicillin" it will be log (1,000,000/10) = log (100,000) = 5.

Hence, the final values for "medicine" and "penicillin" will be 0.05×1 = 0.05 and 0.02×5 = 0.1, respectively.

In the preceding example, we can clearly deduce that using tf-idf the relative importance of "penicillin" for that document has been identified. This is the precise advantage of tf-idf; it reduces the impact of tokens that occur quite frequently. Such tokens which have higher frequency might not offer any information as compared to words which are rare but carry more importance and weight.

The next type of representations we want to discuss are n-grams and language models.

N-gram and Language Models

In the last sections we have studied the bag-of-words approach and tf-idf. Now we are focusing on language models. We understand that to analyze the text data they have to be converted to feature vectors. N-gram models help in creating those feature vectors so that text can be represented in a format which can be analyzed further.

Language models assign probabilities to the sequence of words. N-grams are the simplest in language models. In the n-gram model we calculate the probability of the Nth word given the sequence of (N–1) words. This is done by calculating the relative frequency of the sequence occurring in the text corpus. If the items are words, n-grams may be referred to as *shingles*. Hence, if we have a unigram it is a sequence of one word, for two words it is bi-gram, for three words it is tri-gram, and so on. Let us study by means of an example.

Consider we have a sentence, "Machine learning is very interesting." This sentence can be represented using N=1, N=2, and N=3. You should note how the sequence of words and their respective combinations are getting changed for different values of N, as shown in Figure 4-17.

So, a tri-gram model will approximate the probability of a word given all the previous words by using the conditional probability of only the preceding two words. Whereas a bi-gram will do the same by considering only the preceding word. This is a very strong assumption indeed—that the probability of a word will depend only on the preceding words—and is referred to as a *Markov* assumption.

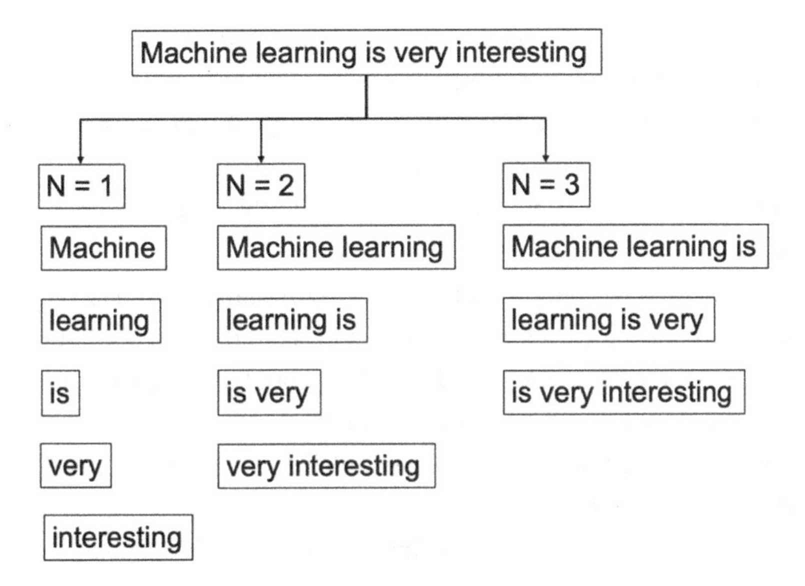

Figure 4-17. *Unigram, bi-gram, tri-gram representation of the same sentence showing different results*

Generally, N > 1 is considered to be much more informative than unigrams. But this approach is very sensitive to the choice of N. It also depends significantly on the training corpus which has been used, which makes the probabilities heavily dependent on the training corpus. So, if we have trained an ML model using a known corpus, we might face difficulties when we encounter an unknown word.

We have studied concepts to clean the text data, tokenize the data, and represent it using multiple techniques. It is time for us to create the first solution in NLP using Python.

Case study: Customer complaints analysis using NLP

In the last section, we examined how to represent text data into feature spaces which can be consumed by an ML model. It is the only difference a text data has from a standard ML model we have created in previous chapters.

In other words, the preprocessing and feature extraction will clean the text data and generate features. The resultant features can then be consumed by any standard supervised learning problem. After the step of feature extraction, a standard ML approach can be followed. We will now solve a case on text data and will create a Python supervised learning algorithm.

Consider we have a dataset of customer complaints. For each customer complaint, we have a corresponding product related to it. We will be using NLP and ML to create a supervised learning model to assign any incoming new complaint to the corresponding product.

The dataset and the code have been uploaded to the Github link shared at the start of the chapter.

Step 1: Import all the necessary libraries and load the dataset.

```
from sklearn.feature_extraction.text import TfidfVectorizer
from sklearn.model_selection import train_test_split
import pandas as pd
complaints_df = pd.read_csv('complaints.csv')
```

Step 2: Let us have a look at a complaint.

```
complaints_df['Consumer complaint narrative'][1]
```

```
complaints_df['Consumer complaint narrative'][1]

"I purchased a new car on XXXX XXXX. The car dealer called Citizens Bank to get a 10 day payoff on my loan, good till
XXXX XXXX. The dealer sent the check the next day. When I balanced my checkbook on XXXX XXXX. I noticed that Citizens
bank had taken the automatic payment out of my checking account at XXXX XXXX XXXX Bank. I called Citizens and they st
ated that they did not close the loan until XXXX XXXX. ( stating that they did not receive the check until XXXX. XXX
X. ). I told them that I did not believe that the check took that long to arrive. XXXX told me a check was issued to
me for the amount overpaid, they deducted additional interest. Today ( XXXX XXXX, ) I called Citizens Bank again and
talked to a supervisor named XXXX, because on XXXX XXXX. I received a letter that the loan had been paid in full ( da
ted XXXX, XXXX ) but no refund check was included. XXXX stated that they hold any over payment for 10 business days a
fter the loan was satisfied and that my check would be mailed out on Wed. the XX/XX/XXXX.. I questioned her about the
delay in posting the dealer payment and she first stated that sometimes it takes 3 or 4 business days to post, then s
he said they did not receive the check till XXXX XXXX I again told her that I did not believe this and asked where is
my money. She then stated that they hold the over payment for 10 business days. I asked her why, and she simply said
that is their policy. I asked her if I would receive interest on my money and she stated no. I believe that Citizens
bank is deliberately delaying the posting of payment and the return of consumer 's money to make additional interest
for the bank. If this is not illegal it should be, it does hurt the consumer and is not ethical. My amount of money l
ost is minimal but if they are doing this on thousands of car loans a month, then the additional interest earned for
them could be staggering. I still have another car loan from Citizens Bank and I am afraid when I trade that car in a
nother year I will run into the same problem again."
```

Step 3: Now we will find out the respective categories for a complaint.

```
print(complaints_df.Product.unique())
```

```
print(complaints_df.Product.unique())

['Credit reporting' 'Consumer Loan' 'Debt collection' 'Mortgage'
 'Credit card' 'Other financial service' 'Bank account or service'
 'Student loan' 'Money transfers' 'Payday loan' 'Prepaid card'
 'Virtual currency'
 'Credit reporting, credit repair services, or other personal consumer reports'
 'Credit card or prepaid card' 'Checking or savings account'
 'Payday loan, title loan, or personal loan'
 'Money transfer, virtual currency, or money service'
 'Vehicle loan or lease']
```

Step 4: Next divide the data into training and testing.

```
X_train, X_test, y_train, y_test = train_test_split(
    complaints_df['Consumer complaint narrative'].values,
    complaints_df['Product'].values,
    test_size=0.15, random_state=0)
```

Step 5: Next we calculate the tf-idf scores for each of the unique tokens in the dataset.

```
vectorizer = TfidfVectorizer()
vectorizer.fit(X_train)
X_train = vectorizer.transform(X_train)
X_test = vectorizer.transform(X_test)
X_train, X_test
```

Step 6: We will select the most significant features now.

```
from sklearn.feature_selection import SelectKBest, chi2
ch2 = SelectKBest(chi2, k=5000)
X_train = ch2.fit_transform(X_train, y_train)
X_test = ch2.transform(X_test)
X_train, X_test
```

Step 7: Fit a naïve Bayes model now.

```
from sklearn.naive_bayes import MultinomialNB
from sklearn.metrics import accuracy_score
clf = MultinomialNB()
clf.fit(X_train, y_train)
pred = clf.predict(X_test)
```

Step 8: Print the predictions in the next step.

```
print(accuracy_score(y_test, pred))
0.7656024029369229
```

The accuracy of the model is 76.56%. This standard model can be applied to any supervised classification problem in text analytics. Here we have a multiclass model. It can be scaled down to a binary classification as (Pass/Fail) or to create a sentiment analysis model like (positive, neutral, negative).

We have studied bag-of-words, tf-idf, and N-gram approaches so far. But in all of these techniques, the relationship between words has been neglected. We will now study an important concept which extends the learnings we have in the light of relationships between words—and it is called *word embeddings*.

Word Embeddings

In the last sections, all the techniques discussed ignore the contextual relationship between words. At the same time, the resultant data is very high-dimensional. Word embeddings provide a solution to the problem. They convert the high-dimensional word features into lower dimensions while maintaining the contextual relationship. We can understand the meaning by looking at an example.

In the example in Figure 4-18, the relation of "man" to "woman" is like that of "king" to "queen"; "go" to "going" is like "run" to "running"; and "UK" to "London" is like "Ireland" to "Dublin." This approach considers the context and relationships between the words as compared to frequency-based methods discussed in the last sections, and hence are better suited for text analytical problems.

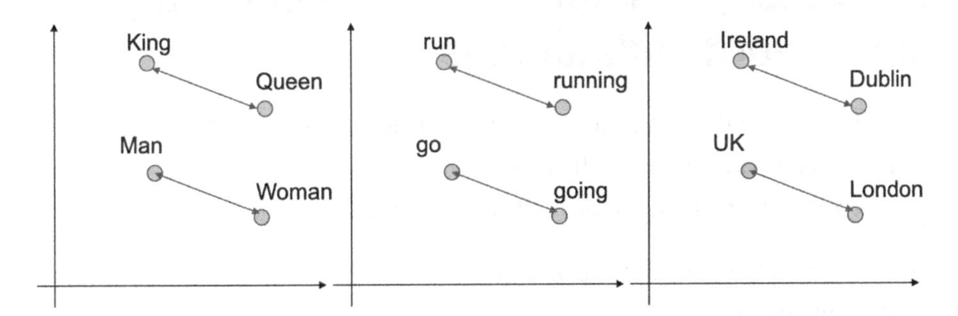

Figure 4-18. *Word embeddings help in finding the contextual relationship between words which are used in the same context and hence improve understanding*

In the example shown previously, the relation of "man" to "woman" is like that of "king" to "queen"; "go" to "going" is like "run" to "running"; and "UK" to "London" is like "Ireland" to "Dublin."

There are two popular word embedding models: Word2Vec and GloVe. *Word2Vec* provides dense embeddings that understand the similarities between "king" and "queen." *GloVe* (Global Vectors for word representations) is an unsupervised algorithm for obtaining representations of words where the training has been performed on aggregated global word-to-word co-occurrence statistics from a corpus.

Both models learn and understand the geometrical encodings or in other words vector representation of their words from the co-occurrence information. Co-occurrence means how frequently the words appear together in the large corpus. The prime difference is that Word2Vec is a prediction-based model, while GloVe is frequency based. Word2Vec predicts the context given a word while GloVe learns the context by creating a co-occurrence matrix on how frequently a word appears in a context. The mathematical details for Word2Vec and GloVe are beyond the scope of this book.

Case study: Customer complaints analysis using word embeddings

We will now use Python and word embeddings to work on the same complaints data we used in the last section.

Step 1: Import the necessary libraries as the first step.

```
from nltk.tokenize import RegexpTokenizer
import numpy as np
import re
```

Step 2: Load the complaints dataset.

```
import pandas as pd
complaints_dataframe = pd.read_csv('complaints.csv')
```

Step 3: Let us now have a look at the first few rows of the data.

```
complaints_dataframe.head()
```

```
complaints_dataframe.head()
```

	Consumer complaint narrative	Product
0	I have outdated information on my credit repor...	Credit reporting
1	I purchased a new car on XXXX XXXX. The car de...	Consumer Loan
2	An account on my credit report has a mistaken ...	Credit reporting
3	This company refuses to provide me verificatio...	Debt collection
4	This complaint is in regards to Square Two Fin...	Debt collection

Step 4: A function has been defined to tokenize the words.

```
def convert_complaint_to_words(comp):

    converted_words = RegexpTokenizer('\w+').tokenize(comp)
    converted_words = [re.sub(r'([xx]+)|([XX]+)|(\d+)', '',
    w).lower() for w in converted_words]
    converted_words = list(filter(lambda a: a != '',
    converted_words))

    return converted_words
```

Step 5: Now we will extract all the unique words from the dataset.

```
all_words = list()
for comp in complaints_dataframe['Consumer complaint
narrative']:
    for w in convert_complaint_to_words(comp):
        all_words.append(w)
```

Step 6: We will now have a look at the size of the vocabulary.

```
print('Size of the vocabulary is {}'.format(len(set(all_words))))
76908
```

Step 7: Print the complaints and the tokens generated.

```
print('Complaint is \n', complaints_dataframe['Consumer
complaint narrative'][10], '\n')
print('Tokens are\n', convert_complaint_to_words (complaints_
dataframe['Consumer complaint narrative'][10]))
```

Step 8: Now we index each unique word in the dataset by assigning it a unique number.

```
index_dictionary = dict()
count = 1
index_dictionary['<unk>'] = 0
```

```
for word in set(all_words):
    index_dictionary[word] = count
    count += 1
```

Step 9: In the next step, indexed words are used to replace words by index, to make the dataset numeric and keras readable.

```
embeddings_index = {}
f = open('glove.6B.300d.txt')
for line in f:
    values = line.split()
    word = values[0]
    coefs = np.asarray(values[1:], dtype='float32')
    embeddings_index[word] = coefs
f.close()
```

Step 10: Now we take an average of all the word embeddings in a sentence to generate the sentence representation.

```
complaints_list = list()
for comp in complaints_dataframe['Consumer complaint
narrative']:
    sentence = np.zeros(300)
    count = 0
    for w in convert_complaint_to_words (comp):
        try:
            sentence += embeddings_index[w]
            count += 1
        except KeyError:
            continue
    complaints_list.append(sentence / count)
```

Step 11: In this step, we convert categorical variables to numeric ones and then one-hot encode them.

```
from sklearn import preprocessing
le = preprocessing.LabelEncoder()
le.fit(complaints_dataframe['Product'])
complaints_dataframe['Target'] = le.transform(complaints_
dataframe['Product'])
complaints_dataframe.head()
```

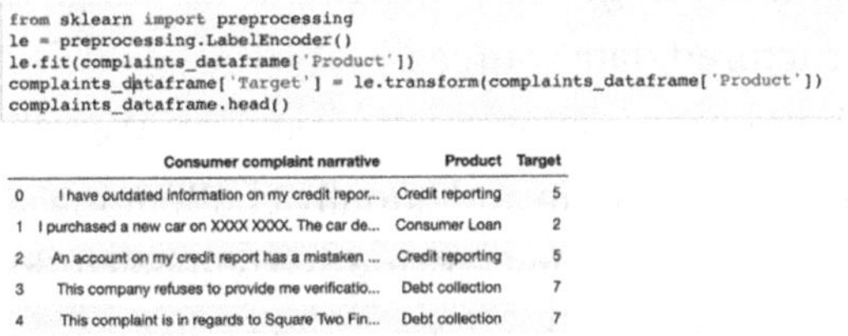

```
from sklearn import preprocessing
le = preprocessing.LabelEncoder()
le.fit(complaints_dataframe['Product'])
complaints_dataframe['Target'] = le.transform(complaints_dataframe['Product'])
complaints_dataframe.head()
```


	Consumer complaint narrative	Product	Target
0	I have outdated information on my credit repor...	Credit reporting	5
1	I purchased a new car on XXXX XXXX. The car de...	Consumer Loan	2
2	An account on my credit report has a mistaken ...	Credit reporting	5
3	This company refuses to provide me verificatio...	Debt collection	7
4	This complaint is in regards to Square Two Fin...	Debt collection	7

```
from sklearn.model_selection import train_test_split
X_train, X_test, y_train, y_test = train_test_split(np.
array(complaints_list), complaints_dataframe.Target.values,
    test_size=0.15, random_state=0)
```

Step 12: Train and test the classifier now.

```
from sklearn.naive_bayes import BernoulliNB
from sklearn.metrics import accuracy_score
clf = BernoulliNB()
clf.fit(X_train, y_train)
pred = clf.predict(X_test)
print(accuracy_score(y_test, pred))
```

In the preceding example, we have used word embeddings to create a supervised classification algorithm. This is a very standard and robust process which can be implemented for similar datasets. For accuracy in a text data, data preprocessing holds the key; the cleaner the data is, the better is the algorithm!

Text data is one of the most interesting datasets to work upon. They are not easy to clean and often require a huge investment of time and processing power to create a model. But despite that, text holds the key to very insightful patterns present in the data. We can use text data for multiple use cases and can generate insights which might not be possible from standard structured data sources.

This concludes our discussion on the text data. We will now move to images, which are as interesting and equally challenging. Since images mostly perform better with deep learning, we will be studying building blocks of neural networks to solve supervised learning case studies for images.

Image Data

If the power of conversation is a gift, vision is a boon to us. We see, we observe, we remember, and we recall whatever we have seen. Through our power of vision, we create a world of images. Images are everywhere. Using our cameras and phones, we click photos. We view photos on social media and at online marketplaces. Images are changing the experience we have, the way we shop, the way we communicate, and the way a business can get its customers.

Similar to test data, images can fall under unstructured data categories. An image is made up of pixels. For each colored image, each pixel can have RGB (red, green, blue) values which range from 0 to 255. We can hence represent each image in pixel values (i.e., in the form of matrix) and do the necessary computations on them.

For illustration purposes in Figure 4-19, we show how an image can be represented in a matrix. These numbers are for illustration only but should give an idea of how we can represent an image in data form, which will be further used for analysis and model building.

11	15	1	20	22
5	12	5	15	21
4	11	25	20	20
6	14	2	21	9
11	21	1	25	21

Figure 4-19. *Illustration to show how can an image be represented in a matrix; the numbers are shown only as an example and are not necessarily correct*

Images are a very powerful source of information. Image data can be analyzed and used for multiple business use cases, which we will discuss next.

Use Cases of Image Data

Consider this. You want to get a coffee from a coffee vending machine. You go to the machine, and the machine recognizes you, recalls your preferences, and delivers precisely what you wanted. The coffee vending machine has recognized your face and based on your previous transactions has given you the desired flavor. Or the attendance monitoring system at an office uses a facial recognition system to mark attendance instead of swiping cards. Analysis of image data and computer vision are enhancing the capabilities and automating processes everywhere.

Image analysis is making ripples across the domains and processes of business. Here are a few use cases:

1. **Healthcare**: Image analysis allows us to identify tumors and illnesses from the x-rays, MRIs, and CT scans. The trained ML model can identify if the image is good or bad, which means are there any signs of illness? The solutions can then locate the problem and then the doctors and medical professionals can use the insights generated and focus on the issues. Patients in remote locations can have their images shared with experts and can get a faster response. We use image segmentation and image classification techniques to identify anomalies and perform analysis.

2. **Retail** industry is harnessing image analysis techniques in a novel way. Customers can upload their pictures of their preferred products like watches, T-shirts, glasses, and so on online and can get a recommendation from the engine. In the background, the online engine will search for similar products and show them to the customer. Moreover, inventory management becomes a lot easier with better image detection techniques. Image segmentation, classification, and computer vision techniques solve the purpose for us.

3. **Manufacturing**: The manufacturing sector employs image analysis techniques in a number of ways:

    a. **Defect identification** is done to separate faulty products from the good ones. It can be implemented using computer vision and image classification.

b. **Predictive maintenance** gets improved by identifying the tools and systems which require maintenance. It makes use of image classification and image segmentation techniques.

4. **Security and monitoring**: Computer vision allows direct monitoring using a security camera and prevents thefts and crime. The capabilities help in crowd management and crowd control, passenger activity movement, and so on. Live monitoring using cameras allows the security teams to prevent any mishaps. Facial recognition techniques allow us to achieve the results.

5. **Agriculture**: The field of agriculture is no different. It also makes use of image recognition and classification techniques to identify weeds amid the plants or any type of disease and infection on the plantations. Soil quality can be checked, and the quality of the grain can be improved. Computer vision is really changing the face of one of the oldest occupations of mankind.

6. **Insurance**: Image analysis is helping the insurance industry by inspecting images from accident sites. Assessment can be made of the damage based on the images from the accident site and claims can be assessed. Image segmentation and image classification drive the solution for the insurance industry.

7. **Self-driving cars** are a very good example of harnessing the power of object detection. Cars, pedestrians, trucks, signs, and so on can be detected and appropriate action can be taken.

8. **Social media platforms and online marketplaces** employ sophisticated image recognition techniques to identify the face, features, expressions, product, and so on based on the photos of the user. It helps them to improve the consumer experience, and improve speed and ease of access.

The preceding use cases are only a few of many use cases where image recognition, object detection, image tracking, image classification, and so on are generating out-of-the-box solutions for us. It is propelled by the latest technology stack of neural networks, convolutional neural networks, recurrent neural networks, reinforcement learning, and so on. They push the boundaries as these solutions are capable of processing tons of complex data easily and generating insights from them. We are able to train the algorithms in a much faster way using modern computing resources and cloud-based infrastructure.

But still we have to explore the full potential of images; we have to improve the existing capabilities and enhance the levels of accuracy. A lot of research is going on in this field with several organizations contributing to the advancement of the sector.

Images are a complex dataset to capture and manage. We will now examine the common challenges we face with the images.

Challenges with Image Data

Images are not an easy dataset to handle. It is a complex amalgamation and a very bulky data point to deal with. Similar to any other dataset, images too are messy and require thorough cleaning. Some of the challenges we face with the image dataset are as follows:

1. **Complexity**: An image of a car will look different from different angles. Front pose vs. left pose vs. right pose of the same person might look completely different, which makes identification of a person or an object by a machine a difficult process. This level of complexity makes image data tougher to analyze.

2. **Size of the dataset**: Size of an image is the next challenge we face with image data. An image can easily be in MB, and based on the frequency of generation, the net size of the image dataset can be really huge.

3. **Images are multidimensional** as compared to any structured data. It also changes as per the image color scale. A multicolored image will have three channels (RGB) and it will increase to the number of dimensions further.

4. **Unclean data**: Images are not always clean. While capturing the dataset itself we face multiple issues. Here are a few of them:

    a. Blurred images are created if the images are out of focus.

    b. There can be shadows on the image which makes it unusable.

    c. Image quality depends on the surrounding lights. If the background light changes, an image will change its compositions.

    d. Distortions happen in the image due to multiple factors like camera vibrations, or the corners are cut or there are marks (like thumb impressions) on the lens.

5. **Human variability**: While capturing the image data, human-generated variance results in different datasets for the same type of problem. For example, if we have to capture the images of crops from a field, different people will capture images from different angles and with different camera modes.

Images are a difficult data set to store and process. Particularly, due to their size, the amount of space required is quite high. We now discuss the image data management process, which concentrates on a few such aspects.

Image Data Management Process

We generate images from multiple sources, and we have to have a concrete data management process for the images. A good system will be able to accept the incoming images, store them, and make them accessible for future analysis. The process of image data management will depend on the design of the system: is it a real-time image analysis project or a batch-processing project? The various sources of images are to be staged, cleaned, and finally stored in a place where they can be accessed as shown in Figure 4-20.

For a real-time image monitoring system, the images are fed to an algorithm in real time and decisions will have to be made in real time. For example, consider if we have a number plate-reading system in a parking lot. The car parking images are generated in real time and have to be processed really fast.

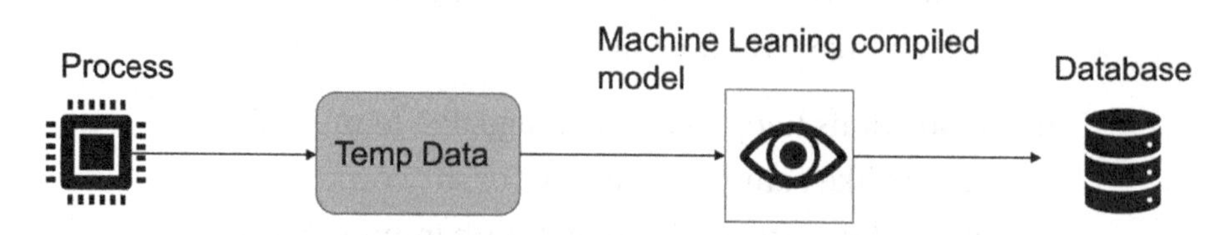

***Figure 4-20.** Real-time process and data management for an ML model*

In the process diagram shown previously, the term "process" represents the source of image data generation. In the number plate reading case discussed previously, it will be the raw image generated by a camera. These raw images might need to be stored temporarily before they are fed to the compiled ML model. The ML model will generate the prediction about the image. In the preceding case, it will be the car registration number. The prediction and the images, both have to then be stored in the final destination database.

For the batch-processing image analysis system, the process changes as shown in Figure 4-21. For example, in the same use case as in the preceding example, if we want to identify how many cars have entered the parking lot in a day, the same image processing system will be different.

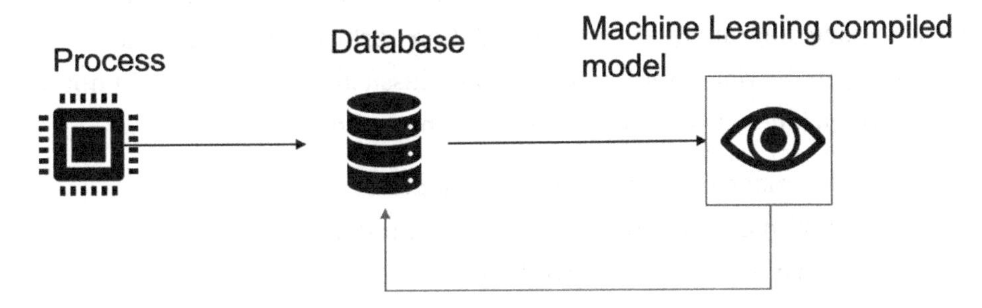

***Figure 4-21.** Batch-processing of image data can receive data from a database, make a prediction, and share the results back to the database to be stored*

For the batch-processing image analysis system, the "process" will generate the images. Those raw images will have to be stored in a database. Then they are fed to the ML model, which generates the predictions for them. In the preceding case, raw images of the cars will be stored as and when they are generated. They are then fed to the image-processing solution as a batch, which generates the registration number for each car. The predictions and the raw images are then sent back to the database and saved.

A good image data management system should be robust, flexible, and easy to access. The size of the images will play a big role in designing the system. It also defines the cost associated with such a database repository. It is worthwhile to note that such a repository will get consumed very soon, so depending on the critical nature of the business and the domain, we might not save all the images. It is also possible that data which is older than a desired frequency might be deleted from the database.

We will now start with the ML modeling process on image data. And for this, we will start with concepts of deep learning in the next section.

Image Data Modeling Process

An image is still a data source. We represent an image as features in a vector-space diagram or as metrics and then perform mathematical modeling on the data. But for images, our classical algorithms might not be able to do justice. This is due to the following reasons:

1. An image dataset has much greater number of dimensions as compared to structured data, and this makes processing difficult.

2. The background noise in case of images is much higher. There can be distortions, blurring, multiple angles shot, gray-scale images, and so on in the image dataset.

3. The size of the input data is again higher than the structured datasets.

Because of a few of the preceding reasons, we prefer to use neural networks to create the image supervised learning algorithms for us, which we discuss in the next section.

Fundamentals of Deep Learning

Deep learning is changing the way we perceive information. It is enhancing the power of data to new levels. Using sophisticated neural networks, we are able to process many complex datasets, which are many dimensional and are of a great size. Neural networks are truly changing the landscape of ML and AI.

Deep learning has created capabilities which were only a thought a few years ago. In the area of image processing, we are implementing neural networks for image classification, object detection, object tracking, image captioning, semantic segmentation, human pose estimation, and so on. The GPUs and TPUs are increasingly allowing us to push the barriers of processing tons of dataset in no time.

In the following section, we are going to discuss the building blocks of neural networks and will be developing a use case in Python.

Artificial Neural Networks

Artificial neural networks or ANNs are arguably inspired by the functioning of a human brain. When we humans see an object for the first time, we create an image of it in our mind and register it. When the same object comes in front of us again, we are able to recognize it easily. The task, which is too easy for us, is quite difficult for algorithms to understand and learn.

Note Depth in deep learning represents the number of hidden layers in the neural network. Generally, the higher the number of hidden layers, the greater is the accuracy. But that is true to a certain extent only, and sometimes the accuracy might not increase even with increasing the number of layers.

We train neural networks like we train any ML algorithm—there is an input dataset, we process it, and the algorithm will generate the output predictions for us. A neural network can be used for both regression and classification problems. It can be used for both structured and unstructured data sources. The levels of accuracy by a neural network are generally higher than a classical ML algorithm like regression, decision tree, and so on. But that might not always be true.

The biggest advantage we have with a neural network is its ability to process complex data like images and videos. Then, recall for classical ML algorithms we choose the significant variables. In the case of neural networks, it is the responsibility of the network to pick the most significant attributes from the data.

A typical neural network looks like Figure 4-22.

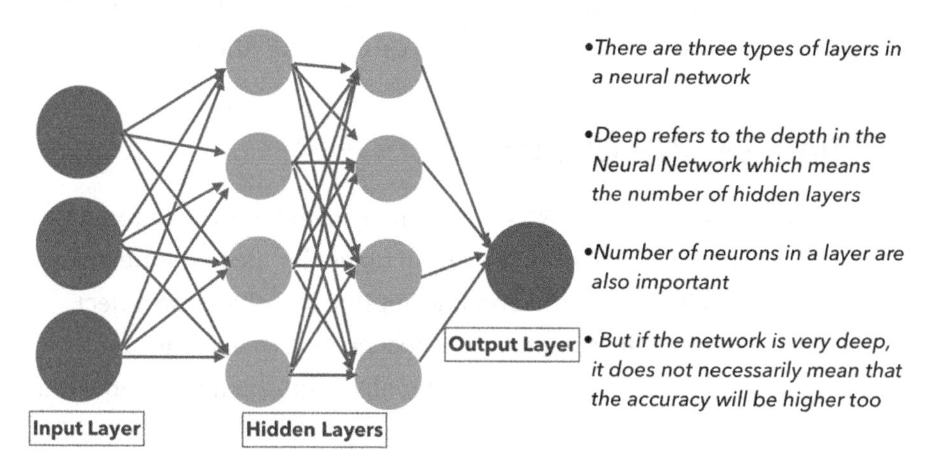

Figure 4-22. *A neural network has input layer, hidden layers, and output layer*

In the structure of the network shown previously, there are few important building blocks which are as follows:

1. **Neuron**: A neuron is a foundation of a neural network. All the calculations and complex processing take place inside a neuron only. It expects an input data like image and will generate an output. That output might be consumed by the next layer in the network or might be used to generate the final result. A neuron can be represented as shown in Figure 4-23. Here, x_0, x_1, and x_2 represent the input variables, and w_0, w_1, and w_2 are their respective weights. "f" is the activation function and "b" is the bias term.

   A neuron receives input from the previous layers and then based on the conditions set, decides whether it should *fire* or not. Simply put, a neuron will receive an input, perform a mathematical calculation on it, and then based on the threshold set inside itself will pass on the value to the next neuron.

   During the training of the ML model or the network in this case, weights and bias terms get trained and we get the most optimized value. We will be discussing the training mechanism and activation term in the next section.

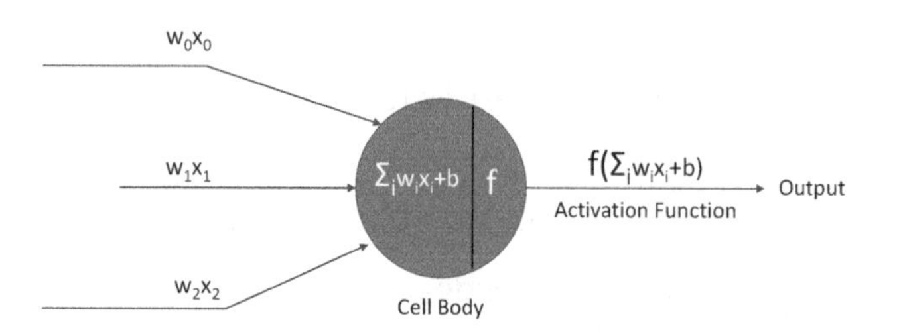

***Figure 4-23.** Basic structure of a neuron showing inputs, weights, activation function, and an output*

2. **Input Layer**: As the name signifies, input layer accepts the input data. It can be in the form of images (raw or processed). This input layer is the first step in the network.

3. **Hidden Layer**: Hidden layers are the most important segment in a neural network. All the complex processes and mathematical calculations take place in the hidden layer only. They accept the data from the input layer, process it layer by layer, and then feed it to the output layer.

4. **Output Layer**: Output layer is the last layer in a neural network. It is responsible for generating the prediction, which can be a continuous variable for a regression problem or a probability score for a supervised classification problem.

With these building blocks of a neural network, in the next section we will discuss the other core elements of a network, which are activation functions.

Activation Functions

Activation functions play a central role in training of a neural network. An activation function's primary job is to decide whether a neuron should fire or not. It is the function which is responsible for the calculations which take place inside a neuron.

The activation functions are generally nonlinear in nature. This property of theirs allows the network to learn complex behaviors and patterns.

There are many types of activation functions available to be used, such as the following:

1. **Sigmoid activation function**: Thist is a bounded mathematical function as shown in Figure 4-24. The range of a sigmoid function is between 0 and 1. The function is S shaped and has a non-negative derivative function.

   Mathematically, a sigmoid function is

$$S(x) = \frac{1}{1+e^{-x}} = \frac{e^x}{e^x+1} \qquad \text{(Equation 4-2)}$$

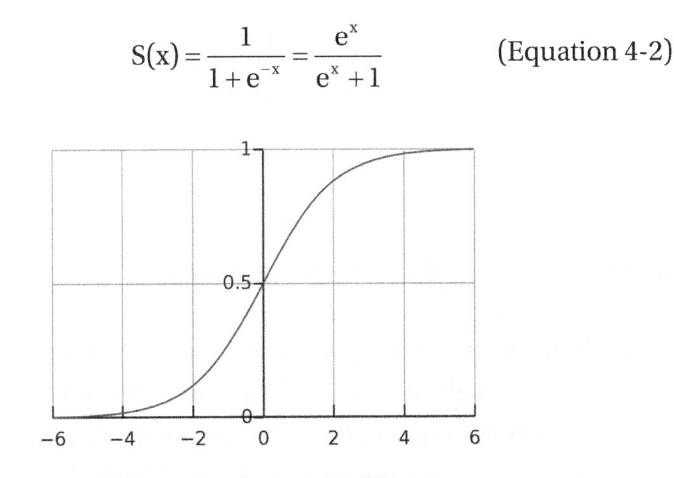

Figure 4-24. *A sigmoid function has an S-like shape*

   It is usually used for binary classifications and in the final output layer of the neural network. But it can be used in the hidden layers of the network too.

2. **tanh activation function**: Tangent hyperbolic function or tanh is a scaled version of the sigmoid function as visible in Figure 4-25. As compared to the sigmoid function, tanh is zero centered. The value ranges between -1 and +1 for tanh function.

   Mathematically, tanh function is given by

$$\tanh = \frac{\left(e^{x} - e^{-x}\right)}{\left(e^{x} + e^{-x}\right)} \qquad \text{(Equation 4-3)}$$

   tanh activation function is generally used in the hidden layers of the neural network. It makes the mean closer to zero, which makes the training easier for the network.

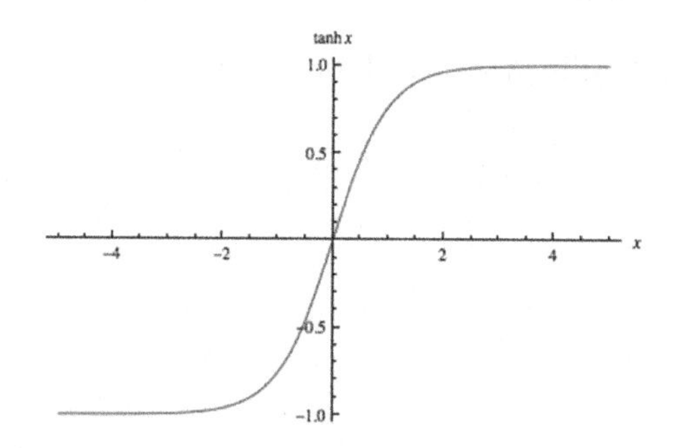

Figure 4-25. A tanh function is centered at zero

3. **ReLU activation function**: Perhaps the most popular of the activation functions is the ReLU activation function. ReLU is a rectified linear unit and is shown in Figure 4-26.

   F(x) = max (x,0) gives the output as x if x>0; otherwise, the output is 0.

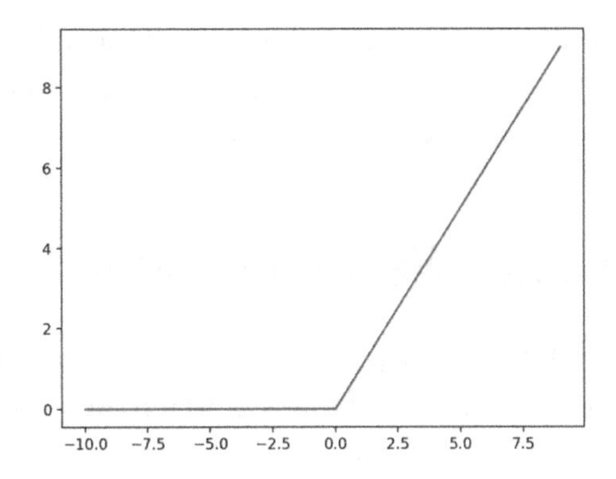

Figure 4-26. *A ReLU function*

ReLU is a very simple function to compute, as visible from the simple mathematical function. It makes ReLU very easy to compute and very fast to train. It is used in the hidden layers of the network.

4. **Softmax function**: The softmax function is used in the final layer of the network used for classification problems. It is used for generating the predictions in the network. The function generates probability scores for each of the target classes, and the class which receives the highest probability is the predicted class. For example, if the network is aimed to distinguish between a cat, dog, a horse, and a tiger, the softmax function will generate four probability scores. The class which receives the highest probability is the predicted class.

Activation functions play a central role in training the network. They define the training progress and are responsible for all the calculations which take place in various layers of the network. A well-designed network will be optimized and then the training of the model will be suitable to be able to make the final predictions. Similar to a classical ML model, a neural network aims to reduce the error in predictions, also known as the loss function, which we cover in the next section.

Loss Function in a Neural Network

We create an ML model to make predictions for the unseen dataset. An ML model is trained on a training dataset, and we then measure the performance on a testing or validation dataset. During measuring the accuracy of the model, we always strive to minimize the error rate. This error is also referred to as *loss*.

Formally put, loss is the difference between the actual values and predicted values by the network. For us to have a robust and accurate network, we always strive to keep this loss to the minimum.

We have different loss functions for regression and classification problems. *Cross-entropy* is the most popular loss function for classification problems, and mean squared error is preferred for regression problems. Different loss functions give a different value for the loss and hence impact the final training of the network. The objective of training is to find the minimum loss, and hence the loss function is also called *objective function*.

Note binary_crossentropy can be used as a loss function for binary classification model.

The neural network is trained to minimize this loss and the process to achieve it is discussed next.

Optimization in a Neural Network

During the training of the neural network, we constantly strive to reduce the loss or error. The loss is calculated by comparing the actual and predicted values. Once we have generated the loss in the first pass, the weights have to be updated to reduce the error further. The direction of this weight update is defined by the *optimization function.*

Formally put, optimization functions allow us to minimize the loss and reach the global minimum. One way to visualize the optimization is as follows: imagine you are standing on top of a mountain. You have to reach the bottom of the mountain. You can take steps in any direction. The direction of the step will be wherever we have the steepest slope. Optimization functions allow us to achieve this. The amount of step we can take in one stride is referred to as the *learning rate.*

We have many choices to use for an optimization function. Here are a few of these choices:

1. **Gradient descent** is one of the most popular optimization functions. It helps us achieve this optimization. Gradient descent optimization is quite fast and easy to implement. We can see gradient descent in Figure 4-27.

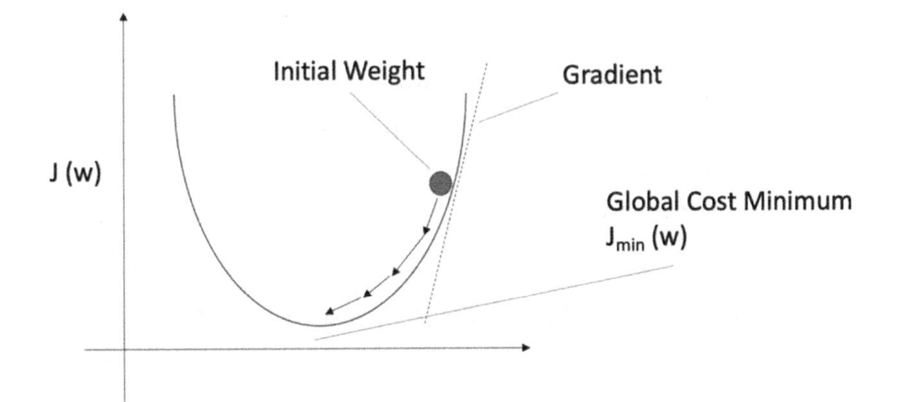

Figure 4-27. *Gradient descent is used to optimize the loss for a neural network*

But gradient descent can be trapped at the local minima and it requires more resources and computation power to be executed.

2. **Stochastic gradient descent** or SGD is a version of gradient descent only. As compared to its parent, it updates the parameters after each training example, that is, after the loss has been calculated after each training example. For example, if the dataset contains 5000 observations, gradient descent will update the weights after all the computations are done and only once, whereas SGD will update the weights 5000 times. While this increases accuracy and decreases computation memory requirements, it results in overfitting of the model too.

3. **Minibatch gradient descent** is an improvement over SGD. It combines best of the gradient descent and SGD. In minibatch gradient descent, instead of updating the parameters after each training example, it updates them in batches. It requires a minimum amount of computation memory and is not prone to overfitting.

4. There are other optimization functions too like Ada, AdaDelta, Adam, Momentum, and so on which can be used. The discussion of these optimizers is beyond the scope of this book.

Note Adam optimizer and SGD can be used for most problems.

Optimization is a very important process in neural network training. It makes us reach the global minima and achieve the maximum accuracy. Hence, due precaution is required while we choose the best optimization function.

There are a few other terms which you should be aware of, before we move to the training of a network.

Hyperparameters

A network learns quite a few parameters itself by analyzing the training examples but a few parameters are required to be fed. Before the training of the neural network commences, we set these parameters to initiate the process. These variables determine the structure of the network, the process, variables, and so on which are used in the final training. They are referred to as *hyperparameters.*

We set a number of parameters like the learning rate, the number of neurons in each layer, activation functions, number of hidden layers, and so on. We also set the number of epochs and batch size. The number of epochs represents the number of times the network will analyze the entire dataset completely. Batch size is the number of samples the network analyzes before updating a model parameter. A batch can contain one or more than one samples.

Simply put, if we have a training data size of 10,000 images, we set batch size as 10 and number of epochs as 50; this means that the entire data will be divided into 10 batches each having 1000 images. The model's weight will be updated after each of those 10 batches. This also means that in one epoch 1000 images will be analyzed 10 times, or in each epoch the weights will be updated 10 times. The entire process will run 50 times as the number of epochs is 50.

Note There are no fixed values of epoch and batch-size. We iterate, measure the loss and then get the best values for the solution.

But training a neural network is a complex process, which we will discuss next. There are processes of forward propagation and backward propagation which are also covered in the next section.

Neural Network Training Process

A neural network is trained to achieve the business problem for which the ML model is being created. It is a tedious process with a lot of iterations. Along with all the layers, neurons, activation functions, loss functions, and so on, the training works in a step-by-step fashion. The objective is to create a network with minimum loss and optimized to generate the best predictions for us.

To design and create deep learning solutions, we have libraries and frameworks. Here are a few of the popular tools which are used for deep learning:

1. **TensorFlow**: It is developed by Google and is one of the most popular frameworks. It can be used with Python, C++, Java, C#, and so on.

2. **Keras**: It is an API-driven framework and is built on top of TensorFlow. It is very simple to use and one of the most recommended libraries to use.

3. **PyTorch**: PyTorch is one of the other popular libraries by Facebook. It is a very great solution for prototyping and cross-platform solutions.

4. **Sonnet**: It is a product by DeepMind and is primarily used for complex neural architectures.

There are many other solutions like MXNet, Swiftm Gluon, Chainer, and so on. We are using Keras to solve the case studies in Python.

Now we will start examining the learning of a neural network. Learning or training in the case of a network refers to finding the best possible values of the weights and the bias terms while keeping an eye on the loss. We strive to achieve the minimum loss after training the entire network.

The major steps while training a neural network are as follows:

Step 1: In the first step, as shown in Figure 4-28, the input data is passed on to the input layer of the network. The input layer is the first layer. The data is fed into a format acceptable by the network. For example, if the network expects an input image of size 25×25, that image data is changed to that shape and is fed to the input layer.

Next, the data is passed and fed to the next layer or the first hidden layer of the network. This hidden layer will transform the data as per the activation functions associated with that particular layer. It is then fed to the next hidden layer and the process continues.

In the diagram shown in Figure 4-28, we have two hidden layers, and each layer has some weights associated with it. Once the respective transformations are done, the final prediction is generated. Recall in the last section, we discussed the softmax layer, which generates the prediction probabilities for us. These predictions are then to be analyzed in the next step.

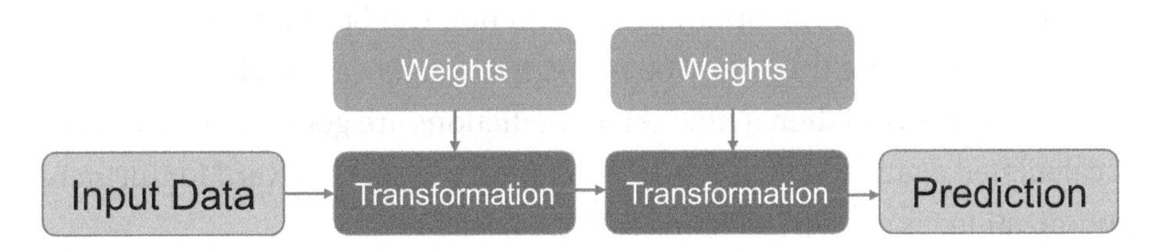

Figure 4-28. *Step 1 in the neural network showing the input data is transformed to generate the predictions*

Step 2: Now we have achieved the predictions from the network. We have to check if these predictions are accurate or not and how far the predicted values are from the actual values. This is done in this step as shown in Figure 4-29.

Here, we compare the actual and predicted values using a loss function and the value of loss is generated in this step.

The feeding of information in this fashion in a forward direction is called a *forward propagation step.*

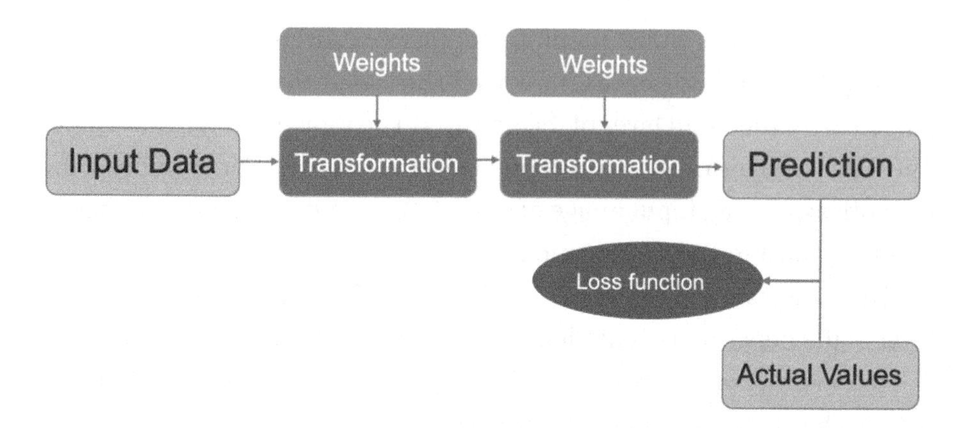

Figure 4-29. *Loss is calculated by comparing the actual and predicted values*

Now we have generated the perceived loss from the first iteration of the network. We still have to optimize and minimize this loss, which is done in the next step of training.

Step 3: In this step, once the loss is calculated, this information travels back to the network. The optimization function will be changing the weights to minimize this loss. The respective weights are updated across all the neurons, and then a new set of predictions are generated. The loss is calculated again and again the information travels backward for further optimization, as shown in Figure 4-30.

This travel of information in a backward direction to optimize the loss is referred to as *backward propagation* in a neural network training process.

Note Backward propagation is sometimes called the central algorithm in deep learning.

This process continues iteratively till we arrive at a point where it is not possible to optimize the loss. And then we can conclude that our network is trained.

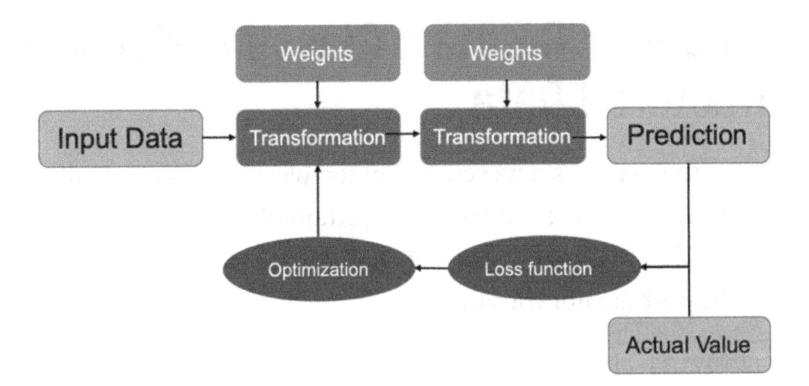

Figure 4-30. *The optimization is done to minimize the loss and hence reach the best solution*

This is the process for a network to train itself and generate a compiled model for us. The training for a neural network is also referred to as *learning* of a network. Formally put, learning of a network refers to finding the most optimal values and best combinations of weights for all layers of the network.

Initially, all the weights are initialized with some random values. The network makes the first prediction, and due to obvious reasons, the loss or the error in the first pass will be quite high. Now, the network encounters new training examples and based on the loss calculated the weights are updated. The backpropagation plays the central role here by acting as a feedback loop. During the process of training the network, these weights are updated during each iteration. The direction of iteration is defined by the gradient of the loss function which allows us to move in the direction to minimize the loss. Once the loss can no longer be decreased, we can say that the network is trained now.

We will now learn how a neural network is trained by creating two use cases in Python: one on structured data and another for images.

Case Study 1: Create a Classification Model on Structured Data

We are using the diabetes data set. The objective is to diagnostically predict if a patient has diabetes or not based on certain diagnostic measurements. The code and dataset are uploaded at Github.

Step 1: Import the libraries first.

```
import numpy as np
import pandas as pd
import seaborn as sns
import matplotlib.pyplot as plt
%matplotlib inline
```

Step 2: Load the data now and have a look at the first five rows:

```
pima_df = pd.read_csv('pima-indians-diabetes.csv')
pima_df.head()
```

```
pima_df.head()
```

	Preg	Plas	Pres	skin	test	mass	pedi	age	class
0	6	148	72	35	0	33.6	0.627	50	1
1	1	85	66	29	0	26.6	0.351	31	0
2	8	183	64	0	0	23.3	0.672	32	1
3	1	89	66	23	94	28.1	0.167	21	0
4	0	137	40	35	168	43.1	2.288	33	1

Step 3: Let's now generate the basic KPI.

```
pima_df.describe()
```

```
pima_df.describe()
```

	Preg	Plas	Pres	skin	test	mass	pedi	age	class
count	768.000000	768.000000	768.000000	768.000000	768.000000	768.000000	768.000000	768.000000	768.000000
mean	3.845052	120.894531	69.105469	20.536458	79.799479	31.992578	0.471876	33.240885	0.348958
std	3.369578	31.972618	19.355807	15.952218	115.244002	7.884160	0.331329	11.760232	0.476951
min	0.000000	0.000000	0.000000	0.000000	0.000000	0.000000	0.078000	21.000000	0.000000
25%	1.000000	99.000000	62.000000	0.000000	0.000000	27.300000	0.243750	24.000000	0.000000
50%	3.000000	117.000000	72.000000	23.000000	30.500000	32.000000	0.372500	29.000000	0.000000
75%	6.000000	140.250000	80.000000	32.000000	127.250000	36.600000	0.626250	41.000000	1.000000
max	17.000000	199.000000	122.000000	99.000000	846.000000	67.100000	2.420000	81.000000	1.000000

Step 4: Plot the data next.

```
sns.pairplot(pima_df, hue='class')
```

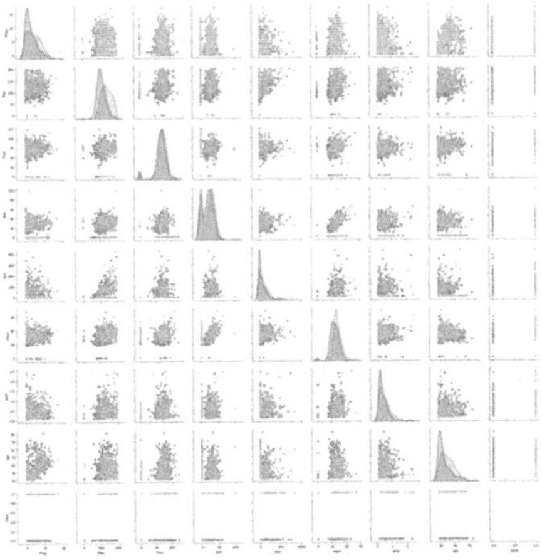

Step 5: We will now generate the correlation plot.

```
sns.heatmap(pima_df.corr(), annot=True)
```

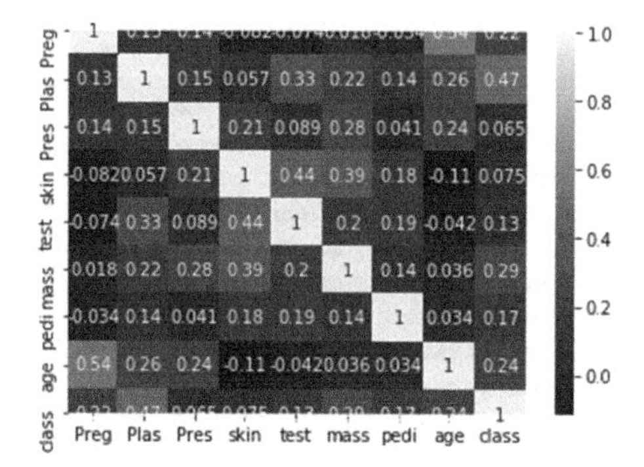

Step 6: Scale the dataset now.

```
X= pima_df.iloc[:,0:8]
y= pima_df.iloc[:,8]
from sklearn.preprocessing import StandardScaler
standard_scaler = StandardScaler()
X = standard_scaler.fit_transform(X)
X
```

Step 7: Split into train and test datasets next.

```
from sklearn.model_selection import train_test_split
X_train, X_test, y_train, y_test = train_test_split(X, y, test_
size=0.2)
```

Step 8: Import the libraries for the neural network creation.

```
from keras import Sequential
from keras.layers import Dense
```

Step 9: Designing of the network begins.

```
diabetes_classifier = Sequential()
```

#First Hidden Layer of the network. In this case, the activation function is ReLU, and the number of neurons is 5. We are initializing the weights as random normal.

```
diabetes_classifier.add(Dense(5, activation='relu', kernel_
initializer='random_normal', input_dim=8))
```

#Second Hidden Layer. Like the last layer, the activation function is ReLU, number of neurons is 5. We are initializing the weights as random normal.

```
diabetes_classifier.add(Dense(5, activation='relu', kernel_
initializer='random_normal'))
```

#Output Layer. The activation function is sigmoid. We are initializing the weights as random normal.

```
diabetes_classifier.add(Dense(1, activation='sigmoid', kernel_
initializer='random_normal'))
```

We are using adam optimizer with cross_entropy as the loss. Accuracy is the metric which has to be optimized.

```
diabetes_classifier.compile(optimizer ='adam',loss='binary_
crossentropy', metrics =['accuracy'])
```

Fit the model now.

```
diabetes_classifier.fit(X_train,y_train, batch_size=10,
epochs=50)
```

```
diabetes_classifier.fit(X_train,y_train, batch_size=10, epochs=50)

Epoch 1/50
614/614 [==============================] - 0s 500us/step - loss: 0.6888 - accuracy: 0.6547
Epoch 2/50
614/614 [==============================] - 0s 127us/step - loss: 0.6758 - accuracy: 0.6547
Epoch 3/50
614/614 [==============================] - 0s 120us/step - loss: 0.6516 - accuracy: 0.6547
Epoch 4/50
614/614 [==============================] - 0s 120us/step - loss: 0.6158 - accuracy: 0.6547
Epoch 5/50
614/614 [==============================] - 0s 98us/step - loss: 0.5786 - accuracy: 0.6547
Epoch 6/50
614/614 [==============================] - 0s 90us/step - loss: 0.5501 - accuracy: 0.6547
Epoch 7/50
614/614 [==============================] - 0s 93us/step - loss: 0.5246 - accuracy: 0.7248
Epoch 8/50
614/614 [==============================] - 0s 87us/step - loss: 0.5015 - accuracy: 0.7818
Epoch 9/50
614/614 [==============================] - 0s 97us/step - loss: 0.4867 - accuracy: 0.7736
Epoch 10/50
614/614 [==============================] - 0s 92us/step - loss: 0.4780 - accuracy: 0.7785
Epoch 11/50
614/614 [==============================] - 0s 96us/step - loss: 0.4732 - accuracy: 0.7801
Epoch 12/50
614/614 [==============================] - 0s 88us/step - loss: 0.4714 - accuracy: 0.7720
Epoch 13/50
614/614 [==============================] - 0s 85us/step - loss: 0.4689 - accuracy: 0.7769
Epoch 14/50
614/614 [==============================] - 0s 88us/step - loss: 0.4672 - accuracy: 0.7769
Epoch 15/50
614/614 [==============================] - 0s 76us/step - loss: 0.4662 - accuracy: 0.7785
Epoch 16/50
614/614 [==============================] - 0s 76us/step - loss: 0.4661 - accuracy: 0.7801
Epoch 17/50
614/614 [==============================] - 0s 84us/step - loss: 0.4649 - accuracy: 0.7850
Epoch 18/50
614/614 [==============================] - 0s 77us/step - loss: 0.4639 - accuracy: 0.7801
Epoch 19/50
614/614 [==============================] - 0s 84us/step - loss: 0.4633 - accuracy: 0.7801
Epoch 20/50
614/614 [==============================] - 0s 76us/step - loss: 0.4626 - accuracy: 0.7834
Epoch 21/50
614/614 [==============================] - 0s 74us/step - loss: 0.4625 - accuracy: 0.7818
Epoch 22/50
614/614 [==============================] - 0s 76us/step - loss: 0.4617 - accuracy: 0.7850
Epoch 23/50
```

Step 10: Check the accuracy of the model using confusion matrix.

```
y_pred=diabetes_classifier.predict(X_test)
y_pred =(y_pred>0.5)
from sklearn.metrics import confusion_matrix
cm = confusion_matrix(y_test, y_pred)
print(cm)
```

```
from sklearn.metrics import confusion_matrix
cm = confusion_matrix(y_test, y_pred)
print(cm)

[[86 12]
 [23 33]]
```

Here we can deduce that the model has an accuracy of 77.27%.

You are advised to test and iterate by

1. Increasing the complexity of the network by adding one or two more layers.
2. Testing with different activation functions. We have used sigmoid; you can use tanh.
3. Increasing the number of epochs and checking the performance.
4. Preprocessing the data better and then again checking the performance over the revised dataset.

Case Study 2: Image Classification Model

We will develop an image classification model on fashion MNIST dataset. It has 70,000 grayscale images in 10 different categories. This dataset is used for image classification problems and is a standard dataset to be used. It comes prebuilt with Keras and can be loaded easily.

Step 1: Import the libraries first.

```
import tensorflow as tf
from tensorflow import keras

import numpy as np
import matplotlib.pyplot as plt
```

Step 2: Import the fashion MNIST dataset which is shipped with Keras and divide into train and test images.

```
fashion_df = keras.datasets.fashion_mnist

(x_train, y_train), (x_test, y_test) = fashion_df.load_data()
```

Step 3: The various groups of apparel available to us are

```
apparel_groups = ['T-shirt/top', 'Trouser', 'Pullover', 'Dress',
'Coat', 'Sandal', 'Shirt', 'Sneaker', 'Bag', 'Ankle boot']
```

Step 4: We will explore the data next.

```
x_train.shape
len(y_train)
```

Step 5: Let's have a look at an element of the data and then preprocess the dataset. We have to standardize the dataset by dividing by 255 (the pixel values range from 0 to 255; hence we are dividing by 255 to standardize the values for all the datasets).

```
plt.figure()
plt.imshow(x_train[1])
plt.show()
```

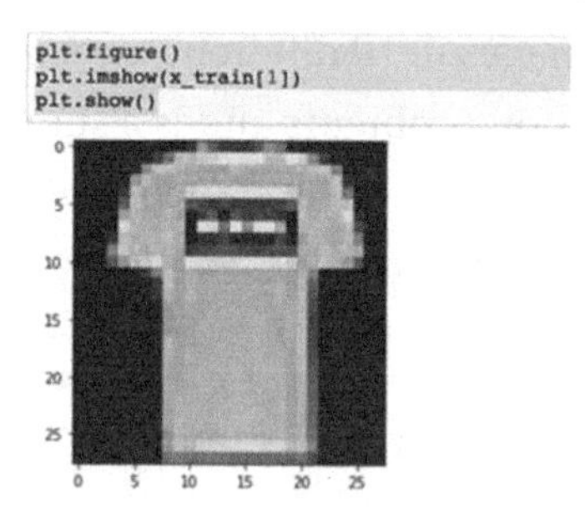


```
x_train = train_images / 255.0

x_test = test_images / 255.0
```

Step 6: Let's have a look at some samples. We will also check if the data is in correct order.

```
plt.figure(figsize=(25,25))
for i in range(10):
    plt.subplot(10,10,i+1)
    plt.xticks([])
    plt.yticks([])
    plt.grid(False)
    plt.imshow(x_train[i], cmap=plt.cm.binary)
    plt.xlabel(apparel_groups[y_train[i]])
plt.show()
```

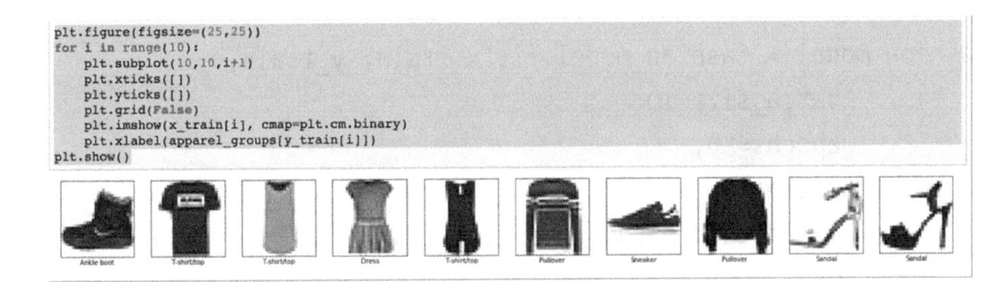

Step 7: We are building the neural network model now.

```
fashion_model = keras.Sequential([
    keras.layers.Flatten(input_shape=(28, 28)),
    keras.layers.Dense(256, activation='relu'),
    keras.layers.Dense(20)
])
```

Here, the input image shape is 28,28. The flatten layer transforms the format of images to 28×28. It only reformats the data and has nothing to learn. Next is the layer that has ReLU as an activation function and 256 neurons. The last layer has 20 neurons and returns a logit array indicating if an image belongs to one of the ten classes we are training upon.

Step 8: Compile the model now.

```
fashion_model.compile(optimizer='adam',
              loss=tf.keras.losses.SparseCategoricalCrossentropy
              (from_logits=True),
              metrics=['accuracy'])
```

The parameters are loss functions to measure how accurate the model is during training. The optimizer determines how the model is updated based on the data it sees and its loss function. We are using accuracy to monitor the training and testing steps.

Step 9: We will train the model now, trying to fit with a batch size of 10 and 50 epochs.

```
fashion_model = fashion_model.fit(x_train, y_train,
          batch_size=10,
          epochs=50,
          verbose=1,
          validation_data=(x_test, y_test))
```

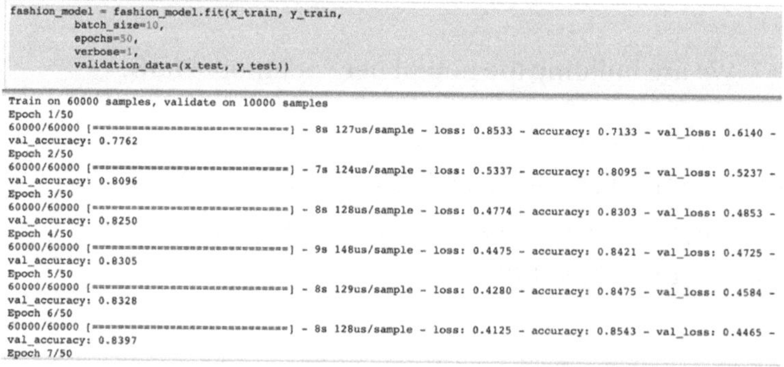

```
fashion_model = fashion_model.fit(x_train, y_train,
          batch_size=10,
          epochs=50,
          verbose=1,
          validation_data=(x_test, y_test))
```

```
Train on 60000 samples, validate on 10000 samples
Epoch 1/50
60000/60000 [==============================] - 8s 127us/sample - loss: 0.8533 - accuracy: 0.7133 - val_loss: 0.6140 -
val_accuracy: 0.7762
Epoch 2/50
60000/60000 [==============================] - 7s 124us/sample - loss: 0.5337 - accuracy: 0.8095 - val_loss: 0.5237 -
val_accuracy: 0.8096
Epoch 3/50
60000/60000 [==============================] - 8s 128us/sample - loss: 0.4774 - accuracy: 0.8303 - val_loss: 0.4853 -
val_accuracy: 0.8250
Epoch 4/50
60000/60000 [==============================] - 9s 148us/sample - loss: 0.4475 - accuracy: 0.8421 - val_loss: 0.4725 -
val_accuracy: 0.8305
Epoch 5/50
60000/60000 [==============================] - 8s 129us/sample - loss: 0.4280 - accuracy: 0.8475 - val_loss: 0.4584 -
val_accuracy: 0.8328
Epoch 6/50
60000/60000 [==============================] - 8s 128us/sample - loss: 0.4125 - accuracy: 0.8543 - val_loss: 0.4465 -
val_accuracy: 0.8397
Epoch 7/50
```

Step 10: We will plot the training and validating accuracy now.

```
import matplotlib.pyplot as plt
f, ax = plt.subplots()
ax.plot([None] + fashion_model.history['accuracy'], 'o-')
```

```
ax.plot([None] + fashion_model.history['val_accuracy'], 'x-')
ax.legend(['Train acc', 'Validation acc'], loc = 0)
ax.set_title('Training/Validation acc per Epoch')
ax.set_xlabel('Epoch')
ax.set_ylabel('acc')
```

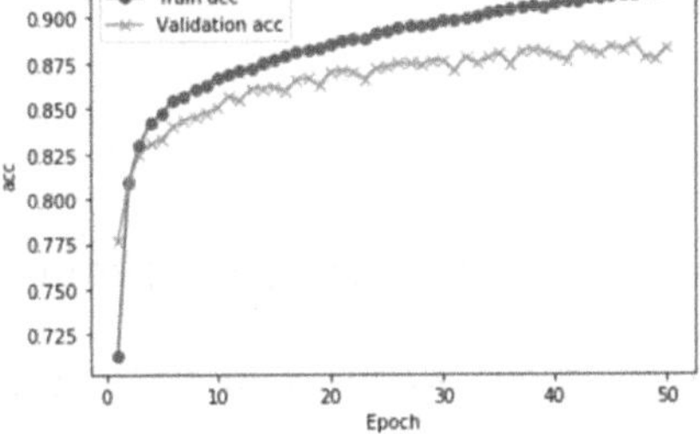

Step 11: We are plotting the training and validating loss.

```
import matplotlib.pyplot as plt
f, ax = plt.subplots()
ax.plot([None] + fashion_model.history['loss'], 'o-')
ax.plot([None] + fashion_model.history['val_loss'], 'x-')
ax.legend(['Train loss', 'Validation loss'], loc = 0)
ax.set_title('Training/Validation loss per Epoch')
ax.set_xlabel('Epoch')
ax.set_ylabel('acc')
```

```
import matplotlib.pyplot as plt
f, ax = plt.subplots()
ax.plot([None] + fashion_model.history['loss'], 'o-')
ax.plot([None] + fashion_model.history['val_loss'], 'x-')
ax.legend(['Train loss', 'Validation loss'], loc = 0)
ax.set_title('Training/Validation loss per Epoch')
ax.set_xlabel('Epoch')
ax.set_ylabel('acc')
```

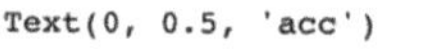

```
Text(0, 0.5, 'acc')
```

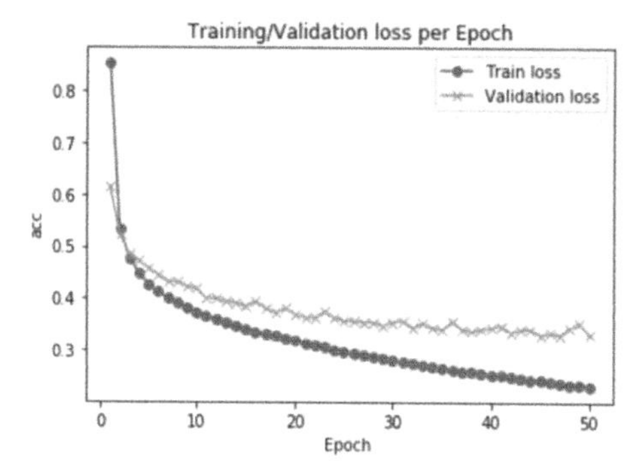

Step 12: We will print the test accuracy we are getting from the model.

```
test_loss, test_acc = fashion_model.model.evaluate(x_test,
y_test, verbose=2)
print('Test accuracy:', test_acc)
```

The test accuracy of the model is 89.03%. Like the last case study, you are advised to iterate by changing the network architecture by adding layers and neurons, and by measuring the performance with different values of epoch, batch_size etc.

We have thus implemented two use cases for structured and unstructured data using deep learning.

Deep learning is the power which allows us to improve capabilities further. For unstructured datasets, neural networks are leading the path in getting solutions which were unheard of before. Advanced neural networks like convolutional neural networks, recurrent neural networks, long short term memory (LSTM), GRU, and so on are doing wonders in every field. We can now detect cancer better, enhance security systems, improve agricultural yields, reduce shopping time, use facial features for allowing or restricting users—the use cases are plenty. And across all the domains, we can feel the ripple effect of deep learning.

With this, we are coming to the end of the fourth chapter of the book. Let's summarize the chapter now.

Summary

This chapter is quite special as it works with advanced concepts. But these advanced concepts are built on top of the basics we have created in the initial chapters. These advanced concepts and algorithms are the need of the hour. With better processing power and enhanced systems, we can implement faster solutions. The required computation power, which is now available, was not in existence a decade back. We now have GPUs and TPUs at our disposal. We now can store and manage terabytes or petabytes of data. Our data collection strategy has improved a lot. We can collect real-time data in a much more structured way. Our data management tools are now not limited to standalone servers only, but have expanded to cloud-based infrastructures. It provides tools and confidence to implement much more advanced ML-based techniques to further improve the capabilities.

At the same time, we cannot undermine the importance of algorithms studied in the previous chapters. Linear regression, decision tree, knn, naïve Bayes, and so on are the foundation of ML algorithms. They are still a preferred choice for shortlisting significant variables. These algorithms set up a benchmark of performance. Ensemble techniques of bagging and boosting then enhance the capabilities further. Hence, before starting with boosting or bagging algorithms, we should always test and benchmark the performance with the base algorithms.

With the advent of deep learning, we are able to process much more complex datasets. Deep learning gives an extra push to the performance. But for processing deep learning algorithms, we require better hardware systems. Cloud-based infrastructures offer good services in this case.

We can deploy our code in Google Colaboratory and run the code, thereby utilizing the processing power of the servers.

With this, we have discussed the supervised learning algorithms within the scope of the book. In Chapter 1, we introduced ML. In Chapters 2 and 3, we examined regression and classification algorithms. In this fourth chapter we studied advanced algorithms like boosting, SVMs, and deep learning models and worked on both structured and unstructured data. In the next chapter, which is the last chapter of the book, we will examine the end-to-end process of a model's life—from scratch to maintenance.

You should now be able to answer the following questions.

EXERCISE QUESTIONS

Question 1: What are different versions of gradient boosting available?

Question 2: How does an SVM algorithm distinguish between classes?

Question 3: What are the data preprocessing steps for text data?

Question 4: What are the various layers in a neural network?

Question 5: What is the function of a loss function and an optimization function in a neural network?

Question 6: Take the datasets used in the last chapter for classification problems and test the accuracy using SVM and boosting algorithms.

Question 7: Download the breast cancer classification data from `https://www.kaggle.com/uciml/breast-cancer-wisconsin-data`. Clean the data and compare performance of random forest and SVM algorithms.

Question 8: Get the Amazon reviews dataset from `https://www.kaggle.com/bittlingmayer/amazonreviews`. Here the problem is to analyze customer reviews on Amazon as input text and output ratings as the output label. Clean the text data using the techniques discussed and create a classification algorithm to make predictions.

Question 9: Download the movie reviews text dataset from `http://ai.stanford.edu/~amaas/data/sentiment/` and create a binary sentiment classification problem.

Question 10: Get the images from `https://www.kaggle.com/c/dogs-vs-cats` and create a binary image classification solution using neural network to distinguish between a dog and a cat.

Question 11: Extend the preceding problem to multiclass problem by downloading the data from `https://www.cs.toronto.edu/~kriz/cifar.html`.

Question 12: Go through the following research papers:

(1) `https://www.sciencedirect.com/science/article/abs/pii/S0893608014002135`

(2) `https://www.sciencedirect.com/science/article/pii/S0893608019303363`

(3) `https://www.aclweb.org/anthology/W02-1011/`

(4) `http://www.jatit.org/volumes/research-papers/Vol12No1/1Vol12No1.pdf`

(5) `https://ieeexplore.ieee.org/document/7493959`

CHAPTER 5

End-to-End Model Development

> *"The journey is the destination."*
>
> — Dan Eldon

The path of learning is never-ending. It takes patience, time, commitment, and hard work. The path is not easy but with commitment a lot can be achieved.

In this journey of learning you have taken the first steps. Data science and ML are changing the business world everywhere. And supervised learning solutions are making an impact across all domains and business processes.

In the first four chapters, we covered regression, classification, and advanced topics like boosting and neural networks. We understood the concepts and developed Python solutions using various case studies. These ML models can be used to make predictions and estimations about various business KPIs like revenue, demand, prices, customer behavior, fraud, and so on.

But in the pragmatic business world, these ML models have to be deployed in a production environment. In the production environment, they are used to make predictions based on real unseen data. We are going to discuss this in this last chapter of the book.

V. Verdhan, *Supervised Learning with Python*,
https://doi.org/10.1007/978-1-4842-6156-9_5

Then there are quite a few best practices with respect to ML and data science. We will be covering all of them. We will also be covering the most common issues faced while creating an ML model like overfitting, null values, data imbalance, outliers, and so on.

Technical Toolkit Required

We are going to use Python 3.5 or above in this chapter. We will be using Jupyter notebook; installing Anaconda-Navigator is required for executing the codes. All the datasets and codes have been uploaded to the Github library at `https://github.com/Apress/supervised-learning-w-python/tree/master/Chapter%205` for easy download and execution.

ML Model Development

Recall that in Chapter 1 we briefly discussed the end-to-end process of model development. We will discuss each of the steps in detail; the most common issues we face with each of them and how we can tackle them are going to be examined in detail now. It will culminate in the model deployment phase. Figure 5-1 is the model development process we follow.

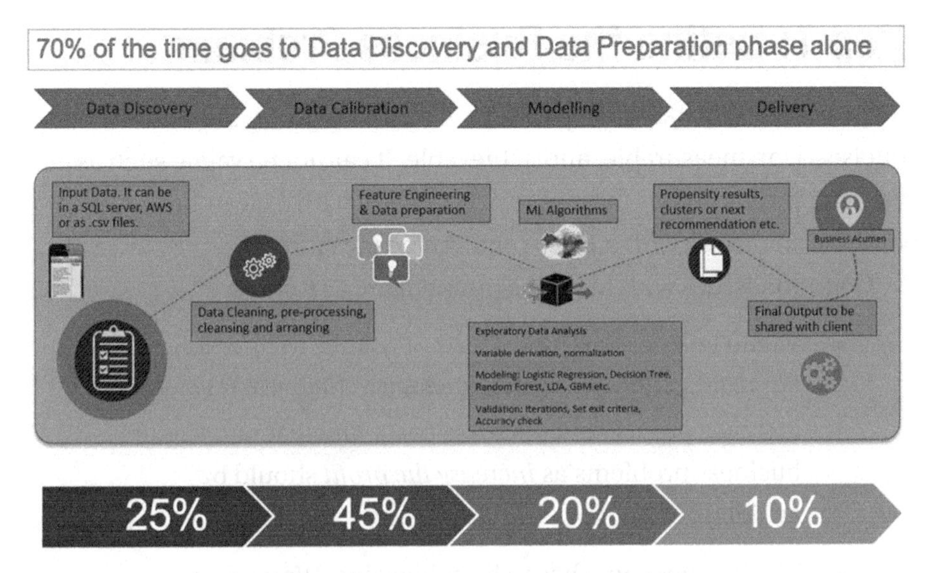

***Figure 5-1.** End-to-end model development process which is generally followed along with respective times spent*

The steps in the ML model development are as follows:

Step 1: Define business problem

Step 2: Data discovery

Step 3: Data cleaning and preparation

Step 4: EDA

Step 5: Statistical modeling

Step 6: Deployment of the model

Step 7: Proper documentation

Step 8: After the model deployment, there is a constant need for model refresh and model maintenance.

We will be discussing all of the steps in detail, the common issues we face in them, and how to tackle them. The entire process will be complemented by actual Python code.

Step 1: Define the Business Problem

It all starts with the business problem. A business problem has to be concise, clear, measurable, and achievable. It cannot be vague such as "increase profits." It has to be precisely defined with a clean business objective and along with a KPI which can be measured.

Common issues with business problem are as follows:

- Vague business problems are a nuisance to be resolved. For example, every business wants to increase revenue, increase profits, and decrease costs. Unclear and vague business problems as *increase the profit* should be avoided.
- The business goal has to be achievable. We cannot expect an ML model to increase the revenue overnight by 80% or decrease costs by half in a month. Hence, the goal has to be achievable within a certain timeframe.
- Sometimes, the business objectives defined are not *quantitative*. If the business problem is to "improve the understanding of the customer base," it does not have a measurable KPI. Hence, it is imperative that a business problem is clearly defined and is measurable by a KPI.
- There is sometimes *scope creep* while deciding on the business problem or during the course of development. Scope creep refers to a situation wherein the initial business objective(s) increase over a period of time or the target changes over time. It also involves change in the business objective. For example, consider a case in which the ML model was envisioned to predict propensity for the customer to churn. Instead, during the course of the project, the objective changes to

predict the propensity of the customer to buy a certain product. It can surely dilute the scope of the solution and will impact the desired output. Such steps should not be encouraged.

A good business problem is one which is precise, achievable, measurable, and repeatable, as shown in Figure 5-2.

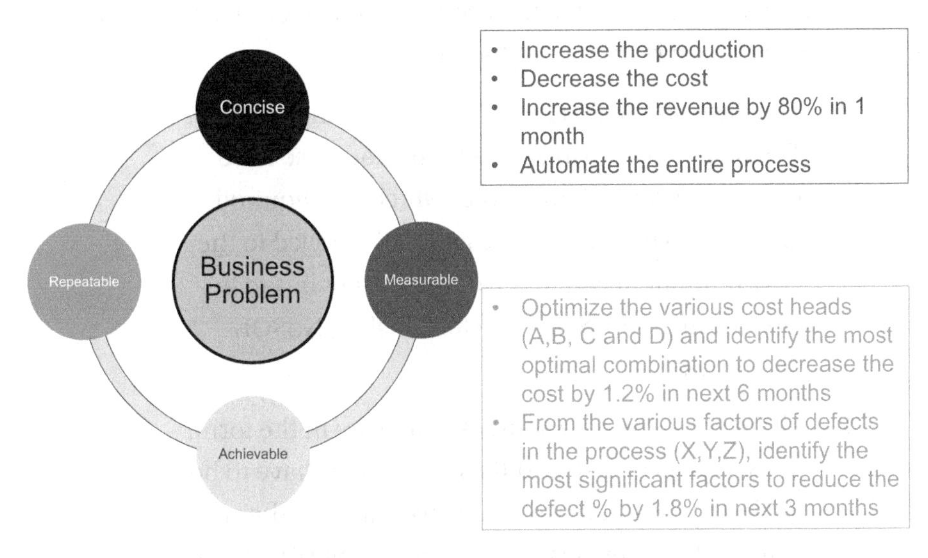

***Figure 5-2.** A good business problem is concise, measurable, achievable and repeatable*

In a nutshell, having a tight scope and a robustly defined business problem paves the way for a successful ML model creation.

Tip Business stakeholders hold the key to the final success of the project. They should be properly engaged in the project from the initial stages.

We now move to the next step, which is the data discovery phase. This step is a very significant one, which will decide if we can go ahead with the ML model or not!

Step 2: Data Discovery Phase

This is the most critical stage of the project, as it decides its future. The data discovery phase includes the following:

1. The database which has to be used for the analysis. The respective permissions to the database have to be requested; all the tables which are required are noted. These tables are generally moved to the environment in which the development has to be done. It might be SQL/Oracle/Redshift/NoSQL databases.

2. There are some data sources which are in the form of Excel files or .csv/text files. These files have to be loaded to the database so that we can load them at will. Moreover, it is a good practice NOT to keep data files on our desktops/laptops. But with the recent surge in data-based architectures, such data sources can be stored on object storage like S3 or Azure Data Lake Storage.

3. The dataset which is to be used has to be complete and relevant for the business problem we want to solve. For example, if we are building a monthly sales estimation model, we should have the data for at least last two or three years with no missing months.

4. The dataset which is used has to be representative enough of the business problem at hand. It should be complete enough so that all the randomness in the business is represented by the dataset. For example, if a manufacturing plant wants to implement a predictive maintenance model, the dataset should contain data for all the equipment which are a part of the business problem.

5. Data refresh also needs to be planned at this stage itself. An ML model might take some time to be built. There might be a need to refresh data before that.

The issues usually faced in the data discovery stage are as follows:

- The dataset is not complete and missing months in between and key information like revenue or number of customers or product information is not available. For example, consider that we are building a model to determine if the customer will churn or not in the coming months. And for that, if we do not have the details of revenue (Table 5-1), then the model will not be able to discover the patterns in the data to predict the churn propensity.

Table 5-1. *Sample Customer Data Having Churn as the Target Variable*

Customer ID	Revenue	# of visits	# of items	Churn (Y/N)
1001	100	2	10	Y
1002	-	3	11	Y
1003	102	4	9	N
1004	?	5	10	N

Hence, in such a case it is imperative to have this missing datapoint for us to make correct deductions about the data.

To resolve such an issue, if we do not get this column then we can make an attempt to estimate the missing data point. For example, in the preceding case, the number of visits can be used as a surrogate for the missing data point.

- A second issue can arise if we have multiple data tables which have data from multiple types of information. If there is a missing link or a missing relationship key, then it will become really difficult to enrich the data. For example, in Table 5-2 we have data from three tables.

Table 5-2. *Tables Representing Employee Data (Employee Demographics, Employment Details, and Salary)*

EmpID	Name	Age	Gender
1	Jack	24	M
2	Tom	25	M
3	Michelle	26	F
4	Lidia	28	F
5	Allan	30	M

EmpID	Grade	Position	Department
1	20	Engineer	Operations
2	21	Engineer	Quality
3	22	Sn Engineer	Process
4	23	Sn Engineer	CRM
5	24	Manager	Marketing

EmpID	Salary	Country	Currency
1	100	US	Dollar
2	101	US	Dollar
3	110	US	Dollar
4	102	Ireland	Euro
5	90	UK	Pound

In Table 5-2, employee ID (EmpID) is the common key across the three tables. If this common relationship is not present or is broken, we cannot join these tables to create a meaningful relationship between all the fields. If there are missing relationships between information in the various tables, we will *not* get a cohesive and complete picture of the dataset.

Unfortunately, if this link is missing, there is not much which can be done. We can still try to join the tables using a surrogate of the department or country, but it might not be 100% accurate.

- Inconsistencies and wrong references to data are problems which can completely flip the analysis and the insights drawn from the data. For example, if there are different systems capturing the customer's revenue and they are not in sync with each other, the results will not be correct.

  The steps to mitigate are as follows:

  a. Create an initial set of reports of the broad KPIs from the dataset: KPIs like revenue, customers, transactions, number of equipment, number of policies, and so on.

  b. Get these KPIs verified and cross-checked with the business stakeholders.

  c. Find the KPIs that have been wrongly calculated and correct them by using the accurate business logic.

  d. Recreate the reports and get a sign-off on the revised results.

- Data quality is one of the biggest challenges. Recall in Chapter 1, we discussed attributes of good-quality data. We will revisit those concepts in detail now and will discuss how to resolve the challenges we face in data in the next step.

- There can be other issues present in the raw data too like the following:

    a. Sometimes, a column might contain information of more than one variable. This can happen during the data capturing phase or the data transformation phase.

    b. The headers of a table are not proper and are numeric. Or they have a space in the names, which can create headaches while we try to refer to these variables in future. For example, Table 5-3 provides two examples of column improper headers in the dataset. The table on the left has numeric column headers and there is a space between the column headers. In the second table, they have been rectified.

Table 5-3. *Improper Headers in the Dataset*

Day Temp	1	2	Humidity	Failed
1	2.5	1.1	100	Y
1	2.1	1.2	100.1	Y
1	2.2	1.2	100.2	Y
1	2.6	1.4	100.3	N
1	2	1.1	99.9	Y
1	2.5	1.1	99.8	N
1	2.1	1.3	100.5	Y
1	2.2	1.3	100.5	Y
1	2.6	1.4	101	Y
1	2	1.1	99	Y

Day_Temp	ABC	PQR	Humidity	Failed
1	2.5	1.1	100	Y
1	2.1	1.2	100.1	Y
1	2.2	1.2	100.2	Y
1	2.6	1.4	100.3	N
1	2	1.1	99.9	Y
1	2.5	1.1	99.8	N
1	2.1	1.3	100.5	Y
1	2.2	1.3	100.5	Y
1	2.6	1.4	101	Y
1	2	1.1	99	Y

In such a case, it is a clever approach to replace the numeric column headers with strings. It is also advised to replace the space with a symbol like "_" or concatenate them. In the preceding example, "Day Temp" has been changed to "Day_Temp" and

"1" and "2" have been changed to "ABC" and "PQR," respectively. In real-world applications, we should find logical names for ABC and PQR.

Data discovery will result in having tables and datasets which are finalized to be used. If there is really a situation where the data is not at all complete, we might decide to not go ahead and complete the dataset instead.

Now it is time to discuss perhaps the most time-consuming of all the steps, which is data cleaning and data preparation.

Step 3: Data Cleaning and Preparation

Perhaps the most-time consuming of all the steps is data cleaning and preparation. It typically takes 60% to 70% of the time along with Step 2.

The data discovery phase will result in creation of datasets which can be used for further analysis. But this does not mean that these datasets are clean and ready to be used. We address them by *feature engineering* and *preprocessing*.

Note Data cleaning is not a fancy job; it requires patience, rigor, and time. It is quite iterative in nature.

The most common issues we face with the data are listed here:

- Duplicates present in the dataset
- Categorical variables are present, but we want to use only numeric features
- Missing data or NULL values or NAN are present in the data
- Imbalanced dataset

- Outliers are present in the data
- Other common problems and transformation done

Let's start with the first issue, which is duplicate values in the dataset.

Duplicates in the Dataset

If we have rows in our dataset which are complete copies of each other, then they are *duplicates*. This means that each value in each of the columns is exactly the same and in the same order too. Such a problem occurs during the data-capturing phase or the data-loading phase. For example, if a survey is getting filled and the same person fills the survey once and then provides exactly the same details again.

An example of duplicate rows is shown in the following Table 5-4. Note that rows 1 and 5 are duplicates.

Table 5-4. *Sample Data with Duplicate Rows That Should Be Removed*

CustID	Revenue	Gender	Items	Date
1001	100	M	4	01-Jan-20
1002	101	F	5	02-Jan-20
1003	102	F	6	04-Jan-20
1004	104	F	8	02-Jan-20
1001	100	M	4	01-Jan-20
1005	105	M	5	05-Jan-20

If we have duplicates in our dataset, we cannot trust the performance of the model. If the duplicated records become a part of both the training and testing datasets, the model's performance will be erroneously elevated and biased. We will conclude that the model is performing well when it is not.

We can identify the presence of duplicates by a few simple commands and drop them from our datasets then.

An example of duplicate rows (row 1 and row 5) is shown in Table 5-4. We will now load a dataset in Python having duplicates and then will treat them. The dataset used is the Iris dataset, which we have used in the exercise sections of Chapters 2 and 3. The code is available at the Github link at `https://github.com/Apress/supervised-learning-w-python/tree/master/Chapter%205`.

Step 1: Import the library.

```
from pandas import read_csv
```

Step 2: Now load the dataset using read_csv command.

```
data_frame = read_csv("IRIS.csv", header=None)
```

Step 3: We will now look at the number of rows and columns in the dataset.

```
print(data_frame.shape)
(151,5) there are 151 rows present in the dataset
```

Step 4: We will now check if there are duplicates present in the dataset.

```
duplicates = data_frame.duplicated()
```

Step 5: We will now print the duplicated rows.

```
print(duplicates.any())
print(data_frame[duplicates])
```

```
True
       0    1    2    3                4
35   4.9  3.1  1.5  0.1      Iris-setosa
38   4.9  3.1  1.5  0.1      Iris-setosa
143  5.8  2.7  5.1  1.9   Iris-virginica
```

From the output we can see that there are three rows which are duplicates and have to be removed.

Step 6: Drop the duplicate rows using the drop_duplicates command and check the shape again.

```
data_frame.drop_duplicates(inplace=True)
print(data_frame.shape)
```

The shape is now reduced to 148 rows.

This is how we can check for the duplicates and treat them by dropping from the dataset set.

Duplicates may increase the perceived accuracy of the ML model and hence have to be dealt with. We now discuss the next common concern: categorical variables.

Categorical Variable Treatment in Dataset

Categorical variables are not exactly a problem in the dataset. They are a rich source of information and insights. Variables like gender, ZIP codes, city, categories, sentiment, and so on are really insightful and useful.

At the same time, having categorical variables in our dataset can pose a few problems like the following:

1. A categorical variable might have a very small number of distinct values. For example, if the variable is "City" and we have a value which is true for a significant percentage of the population, then this variable will be of less use. It will not create any variation in the dataset and will not prove to be useful.

2. Similarly, a categorical variable like "ZIP code" can have a very large number of distinct values. In such a case, it becomes a difficult variable to use and will not add much information to the analysis.

3. ML algorithms are based on mathematical concepts. Algorithms like k-nearest neighbor calculate the distances between data points. If a variable is categorical in nature, none of the distance metrices like Euclidean or cosine will be able to calculate the distances. On one hand, decision trees will be able to handle categorical variables; on the other, algorithms like logistic regression, SVM, and so on do not prefer categorical variables as input.

4. Many libraries like scikit-learn work with numeric data only (as of now). They offer many robust solutions to handle them, some of which we are discussing next.

Hence, it becomes imperative that we treat categorical variables which are present in the dataset. There are quite a few ways to treat the categorical variables. A few of them are given below:

1. Convert the categorical values to numbers. In Table 5-5 we replace the categorical values with numbers. For example, we have replaced the cities with numbers.

Table 5-5. *Categorical Variables Like City Can Be Represented as Numbers*

CustID	Revenue	City	Items
1001	100	New Delhi	4
1002	101	London	5
1003	102	Tokyo	6
1004	104	New Delhi	8
1001	100	New York	4
1005	105	London	5

CustID	Revenue	City	Items
1001	100	1	4
1002	101	2	5
1003	102	3	6
1004	104	1	8
1001	100	4	4
1005	105	2	5

In the example shown, we have replaced the City with a number. But there is a problem in this approach. The ML model will conclude that London is higher ranked than Tokyo or New Delhi is one-fourth of New York. These insights are obviously misleading.

2. Perhaps the most commonly used method is *one-hot encoding*. In one-hot encoding, each of the distinct values is represented as a separate column in the dataset. Wherever the value is present, 1 is assigned to that cell, or else 0 is assigned. The additional variables created are referred to as *dummy variables*. For example, in the preceding dataset after implementing one-hot encoding, the results will look like Table 5-6.

Table 5-6. *One-Hot Encoding to Convert Categorical Variables to Numeric Values, Resulting in an Increase in the Number of Dimensions for the Data*

CustID	Revenue	City	Items
1001	100	New Delhi	4
1002	101	London	5
1003	102	Tokyo	6
1004	104	New Delhi	8
1001	100	New York	4
1005	105	London	5

CustID	Revenue	New Delhi	London	Tokyo	New York	Items
1001	100	1	0	0	0	4
1002	101	0	1	0	0	5
1003	102	0	0	1	0	6
1004	104	1	0	0	0	8
1001	100	0	0	0	1	4
1005	105	0	1	0	0	5

This is surely a much more intuitive and robust method as compared to replacement with numbers. There is a method in pandas as get_dummies which converts categorical variables to dummy variables by adding additional variables. There is one more method in sklearn.preprocessing.OneHotEncoder which can do the job for us. We have already dealt with converting categorical variables to numeric ones in previous chapters of the book.

But one-hot encoding is not a foolproof method. Imagine if we are working on a dataset which has 1000 levels. In that case the number of additional columns will be 1000. Moreover, the dataset will be very sparse as only one column will have "1"; the rest of the values will be 0. In such a case, it is suggested we first combine and club a few categorical values and reduce the number of distinct levels, and then proceed to one-hot encoding.

Categorical variables offer a great deal of insight about the various levels. We have to be cognizant of the algorithms which are capable enough to deal with categorical variables and which are not. It is not always required to convert the categorical variables to numeric ones!

We will now discuss the most common type of defect in the dataset: missing values.

Missing Values Present in the Dataset

Most real-world datasets have missing values for variables. There are NULLS, NAN, 0, blanks, and so on present in the dataset. They can be introduced during the data-capturing part or data transformation, or maybe the data might not be present for those times. Almost all the datasets have missing values, either categorical or numeric. And we have to treat these missing values.

For example, missing values in Table 5-7 have NULL, NAN, and blank values; these missing values have to be treated.

***Table 5-7.** Missing Values Present in the Dataset*

CustID	Revenue	City	Items
1001	NULL	New Delhi	4
1002	101	NULL	5
1003	102	Tokyo	-
1004	NAN	-	8
1001	100	-	NULL
1005	105	London	5

Missing values pose a great problem while we are doing the analysis. It will skew all the means and medians, and we will not be able to draw correct conclusions from the dataset. We will be creating a biased model, as the relationships between the various variables will be biased.

Tip If the value is zero, we cannot conclude that it is a missing value. Sometimes, NULL value is also a correct datapoint. Hence, it is imperative that we apply business logic before treating missing values.

The data can have missing values for the following reasons:

1. During the data extraction phase, the values were not recorded properly. This can be due to faulty equipment or insufficient capabilities to record such data.

2. Many times, the data which is not mandatory is not entered. For example, while filling a form a customer might not want to enter address details.

3. There might be completely random missing values without any pattern or reason.

4. Sometimes (particularly in surveys, etc.), we find responses are not received about a certain attribute of people. For example, responders might not be comfortable in sharing salary details.

5. Missing values might be following a pattern too. For example, we might have data missing for a certain age group or gender or region. This can be due to non-availability or to not being captured for that particular age group, gender, or region.

To mitigate missing values, as a first step, we check if the data is missing by design or if that is an issue which needs to be treated. For example, it is possible for a sensor to NOT record any temperature values above a certain pressure range. In that case, missing values of temperature are correct.

We should also check if there are any patterns in the missing values with respect to the other independent variables and with respect to the target variable. For example, reviewing the dataset in Table 5-8 we can deduce that whenever the value of temperature is NULL, then the equipment has failed. In such a case, there is a clear pattern in this data between temperature and the failed variable. Hence, it will be a wrong step to delete the temperature or treat the temperature variable.

Table 5-8. *Null Values Might Not Always Be Bad*

Temp	Pressure	Viscosity	Humidity	Failed
10	2	1.1	100	N
NULL	2	1.2	100.1	Y
11	1	1.3	100.2	N
12	1	1.4	100.3	N
14	2	1.1	99.9	N
NULL	2	1.1	99.8	Y
11	1	1.2	100.5	N
10	2	1.3	100.5	N
NULL	1	1.4	101	Y
NULL	2	1.1	99	Y

In case we decide that the missing values are to be treated, then there are quite a few ways to do it:

1. Simply **delete** the rows with missing values. It can be considered as the easiest approach where we simply delete the rows which have any value as missing. The biggest advantage is simple and fast implementation. But this approach reduces the size of the population. A blanket deletion of all the rows with missing values can sometimes remove very important pieces of information. Hence, due diligence is required while deletion is used.

2. **Mean, median, or mode imputation**: We might want to impute the missing values by the mean, median, or mode values. Mean and median are only possible for continuous variables. Mode can be used for both continuous and categorical variables.

   Mean imputation or median imputation cannot be implemented without any exploratory analysis of the data. We should understand and analyze if there is any pattern in the data which might get impacted if we impute the missing values with mean or median. For example, in Table 5-9 it will be an incorrect strategy to impute the missing values of temperature with mean since there is a clear indication that temperature has a correlation with pressure and viscosity. If we impute the value with the mean, then it will be a biased dataset. Instead in such a case, we should use a better approach, which is discussed in a later point.

Table 5-9. *Null Values Might Not Always Be Imputed with Mean, Median, and Mode, and Due Analysis Is Required to Reach a Conclusion*

Temp	Pressure	Viscosity	Humidity	Failed
1	2.5	1.1	100	N
2	2.1	1.2	100.1	Y
20	15	20	100.2	N
NULL	16	25	100.3	N
5	2	1.1	99.9	N
NULL	18	28	99.8	Y
21	19	29	100.5	N
2	2	1.3	100.5	N
4	1	1.4	101	Y
5	2	1.1	99	Y

We can use the following Python code to impute the missing values. There is a SimpleImputer class which can be used to impute the missing values as shown in the following code. The code is available at the Github link at `https://github.com/Apress/supervised-learning-w-python/tree/master/Chapter%205`.

```
import numpy as np
from sklearn.impute import SimpleImputer
Next, impute the missing values.
impute = SimpleImputer(missing_values=np.nan,
strategy='mean')
impute.fit([[2, 5], [np.nan, 8], [4, 6]])
SimpleImputer()
X = [[np.nan, 2], [6, np.nan], [7, 6]]
print(impute.transform(X))
```

```
[[3.          2.         ]
 [6.          6.33333333]
 [7.          6.         ]]
```

The output of this function results in the imputation of missing values with mean, as in the preceding function we have used strategy = 'mean'

The same SimpleImputer class can also be used for sparse matrices.

```
import scipy.sparse as sp
matrix = sp.csc_matrix([[2, 4], [0, -2], [6, 2]])
impute = SimpleImputer(missing_values=-1,
strategy='mean')
impute.fit(matrix)
SimpleImputer(missing_values=-1)
matrix_test = sp.csc_matrix([[-1, 2], [6, -1], [7, 6]])
print(impute.transform(matrix_test).toarray())
```

```
[[2.66666667 2.        ]
 [6.         1.33333333]
 [7.         6.        ]]
```

SimpleImputer class can also work on categorical variables where the most frequent value is used for imputation.

```
import pandas as pd
data_frame = pd.DataFrame([["New York", "New
Delhi"],[np.nan, "Tokyo"],["New York", np.nan],[
"New York", "Tokyo"]], dtype="category")
impute = SimpleImputer(strategy="most_frequent")
print(impute.fit_transform(data_frame))
```

```
[['New York' 'New Delhi']
 ['New York' 'Tokyo']
 ['New York' 'Tokyo']
 ['New York' 'Tokyo']]
```

3. **Prediction using an ML model**: We can utilize ML models to estimate the missing values. One of the datasets will be used for training and the other will be one on which prediction has to be made. Then we create the model to make the predictions. For example, we can use k-nearest neighbors to predict the values which are missing. As shown in Figure 5-3, if the value of the missing data point is in yellow, then using knn imputation we can do the missing imputations. We have discussed knn in Chapter 4 in detail. You are advised to go through those concepts again.

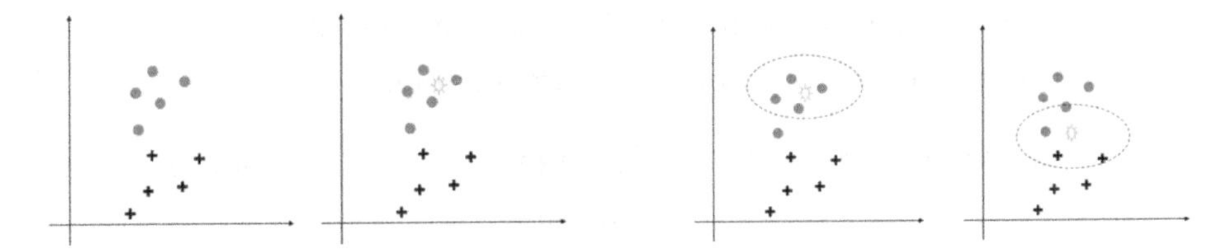

Figure 5-3. *k-nearest neighbor approach used to make imputations to the missing values*

The knn method of imputation is useful for both types of variables: quantitative and qualitative. It considers correlation of various variables, and this is one of the strong advantages of this method. We need not create a predicting algorithm for each of the missing values. It also takes care if there are multiple missing values. But since it searches through the entire dataset for similarities, it becomes quite slow and time-consuming. One of the weaknesses lies in the value of k. This method is quite sensitive to the value of k chosen for the imputation.

Similarly, we can use other algorithms like random forest to impute the missing values. We are developing a code now to generate a sample dataset having missing values and then imputing using knn.

Step 1: Import the necessary libraries.

```
import numpy as np
import pandas as pd
```

Step 2: We will now curate a dataset which contains missing values. We are creating a data frame with five columns, but some values are missing.

```
missing_dictionary = {'Variable_A': [200, 190, 90, 149,
                      np.nan],
                     'Variable_B': [400, np.nan, 149,
                     200, 205],
                     'Variable_C': [200,149, np.nan,
                     155, 165],
                     'Variable_D': [200, np.nan, 90,
                     149,100],
                     'Variable_E': [200, 190, 90, 149,
                     np.nan],}
missing_df = pd.DataFrame(missing_dictionary)
```

Step 3: Let us now have a look at the data frame. We can see that there are a few null values present in the dataset.

```
missing_df
```

	Variable_A	Variable_B	Variable_C	Variable_D	Variable_E
0	200.0	400.0	200.0	200.0	200.0
1	190.0	NaN	149.0	NaN	190.0
2	90.0	149.0	NaN	90.0	90.0
3	149.0	200.0	155.0	149.0	149.0
4	NaN	205.0	165.0	100.0	NaN

Step 4: We will use a knn to impute the missing values. It might require a module KNNImputer to be installed.

```
from sklearn.impute import KNNImputer
missing_imputer = KNNImputer(n_neighbors=2)
imputed_df = missing_imputer.fit_transform(missing_df)
```

Step 5: If we again print the dataset, we can find the missing values have been imputed.

```
imputed_df
```

```
array([[200.  , 400.  , 200.  , 200.  , 200.  ],
       [190.  , 302.5, 149.  , 150.  , 190.  ],
       [ 90.  , 149.  , 160.  ,  90.  ,  90.  ],
       [149.  , 200.  , 155.  , 149.  , 149.  ],
       [169.5, 205.  , 165.  , 100.  , 169.5]])
```

Using a model to predict the missing values serves as a good method. But there is an underlying assumption that there exists a relationship between the missing values and other attributes present in the dataset.

Missing values are one of the most common challenges faced. Seldom will you find a dataset with no missing values. We have to be cognizant of the fact that missing does not necessarily mean incomplete. But once we are sure that there are missing values, it is best to treat them.

We will now move to the next challenge we face, which is an imbalanced dataset.

Imbalance in the Dataset

Consider this. A bank offers credit cards to the customers. Out of all the millions of transactions in a day, there are a few fraudulent transactions too. The bank wishes to create an ML model to be able to detect these fraudulent transactions among the genuine ones.

Now, the number of fraudulent transactions will be very few. Maybe less than 1% of all the transactions will be fraudulent. Hence, the training dataset available to us will not have enough representation from the fraudulent transactions. This is called an *imbalanced* dataset.

The imbalanced dataset can cause serious issues in the prediction model. The ML model will not be able to learn the features of a fraudulent transaction. Moreover it can lead to an *accuracy paradox*. In the preceding case, if our ML models predict all the incoming transactions as genuine, then the overall accuracy of our model will be 99%! Astonishing, right? But the model is predicting all the incoming transactions as genuine, which is incorrect and defeats the purpose of the ML.

There are a few solutions to tackle the imbalance problem:

1. Collect more data and correct the imbalance. This is the best solution which can be implemented and used. When we have a greater number of data points, we can take a data sample having a balanced mix of both the classes in case of a binary model or all the classes in case of a multiclass model.

2. Oversampling and undersampling methods: we can oversample the class which is under-represented OR undersample the class which is over-represented. These methods are very easy to implement and fast to execute. But there is an inherent challenge in both. When we use undersampling or oversampling, the training data might become biased. We might miss some of the important features or data points while we are trying to sample. There are a few rules which we can follow while we oversample or undersample:

    a. We prefer oversample if we have a smaller number of data points.

    b. We prefer undersample if we have a large data set at our disposal.

    c. It is not necessary to have an equal ratio of all the classes. If we have four classes in the target variable, it is not necessary to have 25% of each class. We can target a smaller proportion too.

    d. Use stratified sampling methods while we take the sample (Figure 5-4). The difference between random sampling and stratified sampling is that stratification takes care of distribution of all the variables we want to be used while stratification takes place. In other words, we will get the same distribution of variables in stratified sample as the overall population which might not be guaranteed in random sampling.

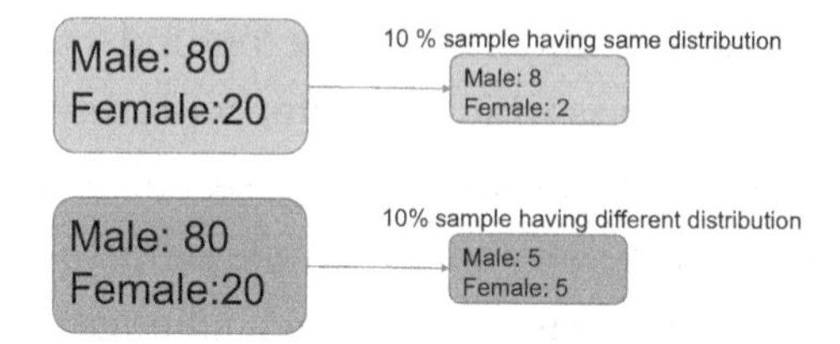

Figure 5-4. *Stratification means that distribution of the variables we care about*

In the population and the sample, stratification maintains the same ratio. For example, as shown in Figure 5-4 the ratio of male to female is 80:20 in the population. In the 10% sample of the population, male to female should be maintained as 80:20, which is achieved by stratification and not necessarily guaranteed by random sampling.

Tip Oversampling/undersampling has to be used only for the training data. The testing data should not be treated. So oversampling/undersampling is done only after creating a train/test split on the data and not before that.

---

3. We can change the way we measure the accuracy of our model. In the example of the fraud detection case study we discussed previously, accuracy will not be the parameter which should be targeted or should be achieved. Instead, we should focus on recall as the parameter which should be optimized and targeted for.

4. We can try a suite of algorithms and not focus on one family or one algorithm alone. This allows us to check and gauge the efficacy of different algorithms. It has been found that decision trees generally work well in such a scenario.

5. We can also implement an SMOTE algorithm which generates or synthesizes data for us. SMOTE is an oversampling technique and it is Systematic Minority Oversampling Technique.

SMOTE works in the following fashion:

a. It analyzes the feature space of the target class.

b. Then it detects the nearest neighbor and selects the data samples which are similar. Generally, two or more instances are selected using a distance metric.

c. It is followed by changing one column at a time randomly within the feature space of the neighbors.

d. Then a synthesized sample is generated which is not the exact copy of the original data.

We can study it by means of an example. The dataset is of a credit card fraud and is imbalanced. We are going to balance it. The dataset and code is available at the Github link at `https://github.com/Apress/supervised-learning-w-python/tree/master/Chapter%205`.

Step 1: Import all the libraries. The SMOTE module might require installation.

```
import pandas as pd

from imblearn.over_sampling import SMOTE

from imblearn.combine import  SMOTETomek
```

Step 2: Read the credit card dataset.

```
credit_card_data_set = pd.read_csv('creditcard.csv')
```

Step 3: We will first create the X and y variables and then resample using the SMOTE method.

```
X = credit_card_data_set.iloc[:,:-1]
y = credit_card_data_set.iloc[:,-1].map({1:'Fraud',
0:'No Fraud'})
X_resampled, y_resampled = SMOTE(sampling_
strategy={"Fraud":500}).fit_resample(X, y)
X_resampled = pd.DataFrame(X_resampled, columns=
X.columns)
```

Step 4: Let us now see what was the % of class in original dataset.

```
class_0_original = len(credit_card_data_set[credit_
card_data_set.Class==0])
class_1_original = len(credit_card_data_set[credit_
card_data_set.Class==1])
print(class_1_original/(class_0_original+class_1_
original))
```

We get the answer as 0.00172 which means 0.172%

Step 5: After resampling, let us analyze the % distribution.

```
sampled_0 = len(y_sampled[y_sampled==0])
sampled_1 = len(y_sampled[y_sampled==1])
print(sampled_1/(sampled_0+sampled_1))
```

We get the answer as 50%. Hence the dataset has been balanced from less than 1% to 50%.

SMOTE is a fantastic method to treat the class imbalance. It is fast and easy to execute and give good results for numeric values. But SMOTE works erroneously for categorical variables. For example, if we have a categorical variable as "Is_raining". This variable can only take value in binary (i.e., 0 or 1). SMOTE can lead to a value in decimals for this variable like 0.55, which is not possible. So be cautious while using SMOTE!

With this, we have completed the imbalance dataset treatment. It is one of the most significant challenges we face: it not only impacts the ML model but has an everlasting impact on the accuracy measurement too. We now discuss the outlier issues we face in our datasets.

Outliers in the Dataset

Consider this. In a business the average sales per day is $1000. One fine day, a customer made a purchase of $5000. This information will skew the entire dataset. Or in a weather forecast dataset, the average rainfall for a city is 100 cm. But due to a drought season there was no rain during that season. This will completely change the face of the deductions we will have from this data. Such data points are called *outliers*.

Outliers impact the ML model adversely. The major issues we face with outliers are as follows:

1. Our model equation takes a serious hit in the case of outliers, as visible in Figure 5-5. In the presence of outliers, the regression equation tries to fit them too and hence the actual equation will not be the best one.

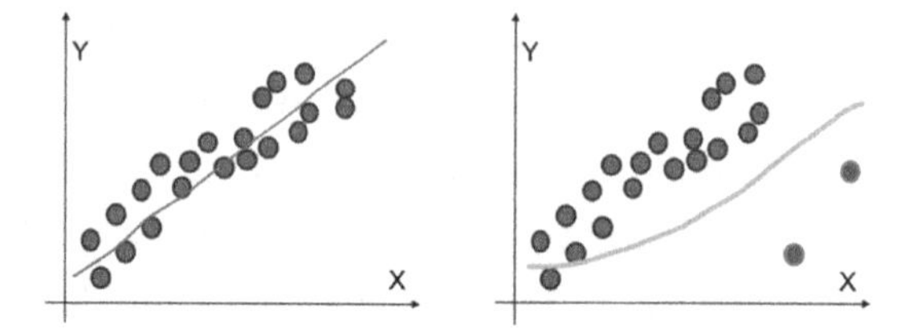

***Figure 5-5.** The impact of outliers on the regression equation makes it biased*

2. Outliers bias the estimates for the model and increase the error variance.
3. If a statistical test is to be performed, its power and impact take a serious hit.
4. Overall, from the data analysis, we cannot trust coefficients of the model and hence all the insights from the model will be erroneous.

But similar to missing values, outliers may not be necessarily bad. We have to apply business logic to check if the value being called as outlier is truly an outlier or is a valid observation. For example, in Table 5-10, the values of temperature look like they are outliers, but once we take a closer look we can deduce that whenever temperature exceeds or falls a lot, the value of the target variable "Failed" is No.

Table 5-10. *Outlier Value of Temperature May Not Necessarily Be Bad and Hence Outliers Should Be Treated with Caution*

Temp	Pressure	Viscosity	Humidity	Failed
1	2.5	1.1	100	Y
2	2.1	1.2	100.1	Y
2	2.2	1.2	100.2	Y
40	2.6	1.4	100.3	N
5	2	1.1	99.9	Y
-20	2.5	1.1	99.8	N
2	2.1	1.3	100.5	Y
5	2.2	1.3	100.5	Y
4	2.6	1.4	101	Y
5	2	1.1	99	Y

Hence, we should be cautious while we treat outliers. Blanket removal of outliers is not recommended!

There are a number of methods to detect outliers, including the following:

1. Recall we discussed normal distribution in Chapter 1. If a value lies beyond the 5th percentile and 95th percentile OR 1st percentile and 99th percentile we may consider it as an outlier.

2. A value which is beyond $-1.5 \times IQR$ and $+1.5 \times IQR$ can be considered as an outlier. Here IQR is interquartile range and is given by (value at 75th percentile) - (value at 25th percentile).

3. Values beyond one or two or three standard deviations from the mean can be termed as outliers.

4. Business logic can also help in determining any unusual value to detect outliers.

5. We can visualize outliers by means of a box-plot or box-whiskers diagram. We have created box-plots in earlier chapters and will be revisiting in the next section.

We can treat outliers using a number of ways. After we are sure that a particular value is an outlier and requires attention, we can do the following:

1. Delete the outliers completely. This is to be done after confirming that outliers can be deleted.
2. Cap the outliers. For example, if we decide that anything beyond 5th percentile and 95th percentile is outlier, then anything beyond those values will be capped at 5th percentile and 95th percentile, respectively.
3. Replace the outliers with mean, median, or mode. The approach is similar to the one we discussed for missing value treatment we discussed in the last section.
4. Sometimes, taking a natural log of the variable reduces the impact of outliers. But again a cautious approach should be taken if we decide to take a natural log as it changes the actual values.

Outliers pose a big challenge to our datasets. They skew the insights we have generated from the data. Our coefficients are skewed and the model gets biased, and hence we should be cognizant of their impact. At the same time, we cannot ignore the relation of the business world with outliers and gauge if the observation is a serious outlier.

Apart from the issues discussed previously, there are quite a few other challenges faced in real-world datasets. We discuss some of these challenges in the next section.

Other Common Problems in the Dataset

We have seen the most common issues we face with our datasets. And we studied how to detect those defects and how to fix them. But real-world datasets are really messy and unclean, and there are many other factors which make a proper analysis fruitful.

Some of the other issues we face in our datasets are as follows:

1. Correlated variables: if the independent variables are correlated with each other, it does not allow us to measure the true capability of our model. It also does not generate correct insights about the predictive power of the independent variables. The best way to understand collinearity is by generating a correlation matrix. This will guide us towards the correlated variables. We should test multiple combinations of input variables and if there is no significant drop in accuracy or pseudo R^2, we can drop one of the correlated variables. We have dealt with the problem of correlated variables in the previous chapters of the book.

2. Sometimes there is only one value present in the entire column. For example, in Table 5-11.

Table 5-11. *Temperature Has Only One Value and Hence Is Not Adding Any Information*

Temp	Pressure	Viscosity	Humidity	Failed
1	2.5	1.1	100	Y
1	2.1	1.2	100.1	Y
1	2.2	1.2	100.2	Y
1	2.6	1.4	100.3	N
1	2	1.1	99.9	Y
1	2.5	1.1	99.8	N
1	2.1	1.3	100.5	Y
1	2.2	1.3	100.5	Y
1	2.6	1.4	101	Y
1	2	1.1	99	Y

Temperature has a constant value of 1. It will not offer any significant information to our model and hence it is our best interest to drop this variable.

3. An extension to the preceding point can be, if a particular variable has only two or three unique values. For example, consider a case when the preceding dataset has 100,000 rows. If the unique values of temperature are only 100 and 101, then it too will not be of much benefit. However, it might be as good as a categorical variable.

Tip Such variables will not add anything to the model and will unnecessarily reduce pseudo R^2.

---

4. Another way to detect such variables is to check their *variance*. The variance of the variables discussed in Point 1 will be 0 and for the ones discussed in Point 2 will be quite low. So, we can check the variance of a variable, and if it is below a certain threshold, we can examine those variables further.

5. It is also imperative that we always ensure that the issues discussed during the data discovery phase in the previous section are tackled. If not, we will have to deal with them now and clean them!

Here are some transformations which can be applied to the variables:

1. Standardization or z-score: here we transform the variable as shown the formula.

$$\text{Standardized x} = \frac{(\text{x} - \text{mean})}{(\text{standard deviation})}$$

   This is the most popular technique. It makes the mean as 0 and standard deviation as 1

2. Normalization is done to scale the variables between 0 and 1 retaining their proportional range to each other. The objective is to have each data point same scale so that each variable is equally important, as shown in Equation 5-1.

$$\text{Normalized x} = \frac{\text{x} - \min(\text{x})}{(\max(\text{x}) - \min(\text{x}))} \qquad \text{(Equation 5-1)}$$

3. Min-max normalization is used to scale the data to a range of [0,1]; it means the maximum value gets transformed to 1 while the minimum gets a value of 0. All the other values in between get a decimal value between 0 and 1, as shown in Equation 5-2.

$$\text{Normalized x} = \frac{(\text{x} - \text{x}_{\text{min}})}{(\text{x}_{\text{max}} - \text{x}_{\text{min}})} \qquad \text{(Equation 5-2)}$$

Since Min-max normalization scales data between 0 and 1, it does not perform very well in case there are outliers present in the dataset. For example, if the data has 100 observations, 99 observations have values between 0 and 10 while 1 value is 50. Since 50 will now be assigned "1," this will skew the entire normalized dataset.

4. Log transformation of variables is done to change the shape of the distribution, particularly for tackling skewness in the dataset.

5. Binning is used to reduce the numeric values to categorize them. For example, age can be categorized as young, middle, and old.

This data-cleaning step is not an easy task. It is really a tedious, iterative process. It does test patience and requires business acumen, dataset understanding, coding skills, and common sense. On this step, however, rests the analysis we do and the quality of the ML model we want to build.

Now, we have reached a stage where we have the dataset with us. It might be cleaned to a certain extent if not 100% clean.

We can now proceed to the EDA stage, which is the next topic.

Step 4: EDA

EDA is one of the most important steps in the ML model-building process. We examine all the variables and understand their patterns, interdependencies, relationships, and trends. Only during the EDA phase will we come to know how the data is expected to behave. We uncover insights and recommendations from the data in this stage. A strong visualization complements the complete EDA.

Tip EDA is the key to success. Sometimes proper EDA might solve the business case.

There is a thin line between EDA and the data-cleaning phase. Apparently, the steps overlap between data preparation, feature engineering, EDA, and data-cleaning phases.

EDA has two main heads: *univariate* analysis and *bivariate* analysis. Univariate, as the name suggests, is for an individual variable. Bivariate is used for understanding the relationship between two variables.

We will now complete a detailed case using Python. You can follow these steps for EDA. For the sake of brevity, we are not pasting all the results here. The dataset and code are available at the Github link at `https://github.com/Apress/supervised-learning-w-python/tree/master/Chapter%205`.

Step 1: Import the necessary libraries.

```
import pandas as pd
import numpy as np
import matplotlib.pyplot as plt
%matplotlib inline
```

Step 2: Load the data file.

```
matches_df = pd.read_csv('matches.csv')
matches_df.head()
```

Now, we will work on the player of the match and visualize it.

```
matches_df.player_of_match.value_counts()
matches_df.player_of_match.value_counts()[:10].plot('bar')
```

```
matches_df.player_of_match.value_counts()

CH Gayle           17
YK Pathan          16
AB de Villiers     15
DA Warner          14
SK Raina           13
                   ..
BCJ Cutting         1
HH Gibbs            1
GD McGrath          1
SM Katich           1
MJ Lumb             1
Name: player_of_match, Length: 187, dtype: int64
```

```
matches_df.player_of_match.value_counts()[:10].plot('ba
```

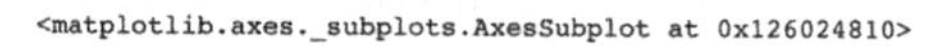

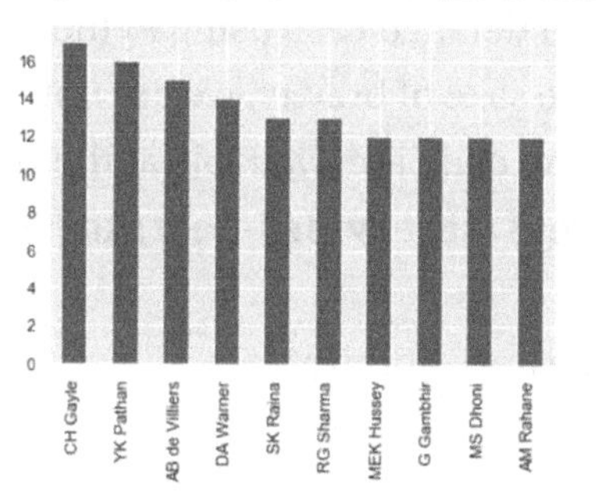

Deliveries dataset is loaded; check the top five rows.

```
deliveries_df = pd.read_csv('deliveries.csv')
deliveries_df.head()
Plot the batsmen by their receptive runs scored
batsman_runs = deliveries_df.groupby(['batsman']).batsman_runs.
sum().nlargest(10)
batsman_runs.plot(title = 'Top Batsmen', rot = 30)
```

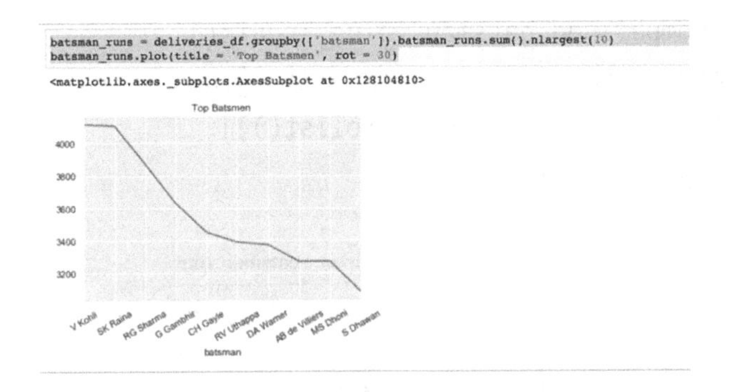

Same plot as a bar plot can be created using this code.

```
deliveries_df.groupby(['batsman']).batsman_runs.sum().
nlargest(10)\
.plot(kind = 'bar',title = 'Top Batsmen', rot = 40, colormap =
'coolwarm')
```

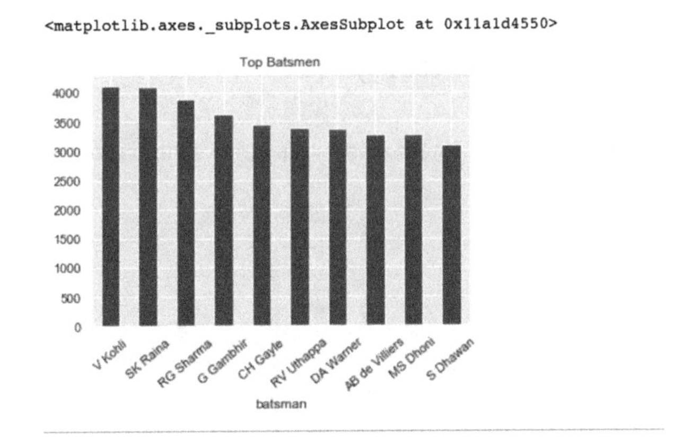

Let us compare the performance of years 2015 and 2016.

```
ipl = matches_df[['id', 'season']].merge(deliveries_df, left_on =
'id', right_on = 'match_id').drop('match_id', axis = 1)
runs_comparison = ipl[ipl.season.isin([2015, 2016])].
groupby(['season', 'batsman']).batsman_runs.sum().nlargest(20).
reset_index().sort_values(by='batsman')
```

```
vc = runs_comparison.batsman.value_counts()
batsmen_comparison_df = runs_comparison[runs_comparison.
batsman.isin(vc[vc == 2].index.tolist())]
batsmen_comparison_df
```

	season	batsman	batsman_runs
2	2016	AB de Villiers	687
6	2015	AB de Villiers	513
4	2015	AM Rahane	540
13	2016	AM Rahane	480
1	2016	DA Warner	848
3	2015	DA Warner	562
11	2016	RG Sharma	489
12	2015	RG Sharma	482
7	2015	V Kohli	505
0	2016	V Kohli	973

```
batsmen_comparison_df.plot.bar()
```

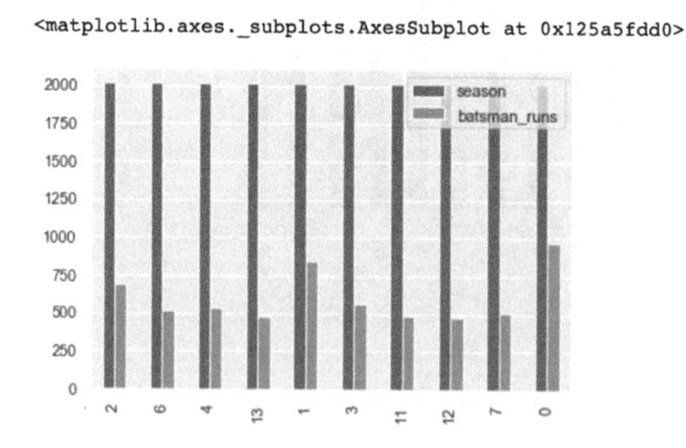


```
batsmen_comparison_df.pivot('batsman', 'season', 'batsman_
runs').plot(kind = 'bar', colormap = 'coolwarm')
```

```
<matplotlib.axes._subplots.AxesSubplot at 0x1260c4b50>
```

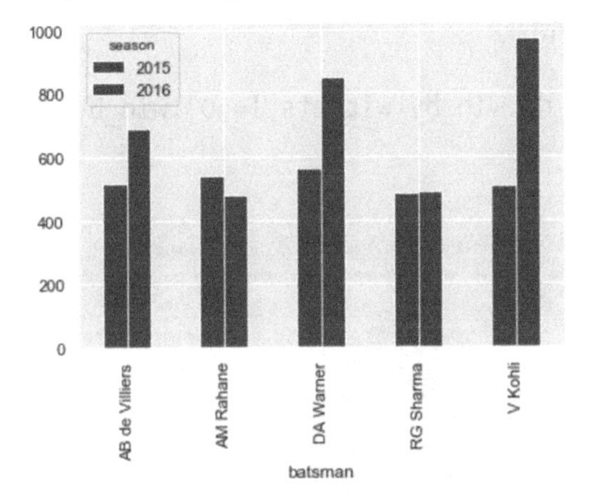

If we want to check the percentage of matches won and visualize it as a pie chart, we can use the following:

```
match_winners = matches_df.winner.value_counts()
fig, ax = plt.subplots(figsize=(8,7))
explode = (0.01,0.02,0.03,0.04,0.05,0.06,0.07,0.08,0.09,0.1,0.2,
0.3,0.4)
ax.pie(match_winners, labels = None, autopct='%1.1f%%',
startangle=90, shadow = True, explode = explode)
ax.legend(bbox_to_anchor=(1,0.5), labels=match_winners.index)
```

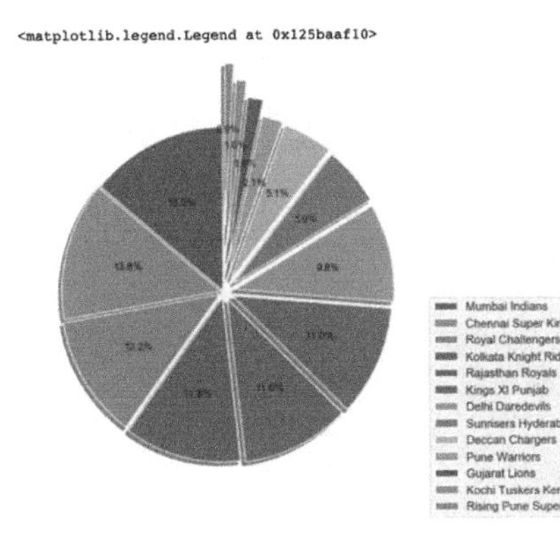

We will now visualize a histogram of dataset where the victory was with wickets and not with runs.

```
matches_df[matches_df.win_by_wickets != 0].win_by_wickets.
hist()
```

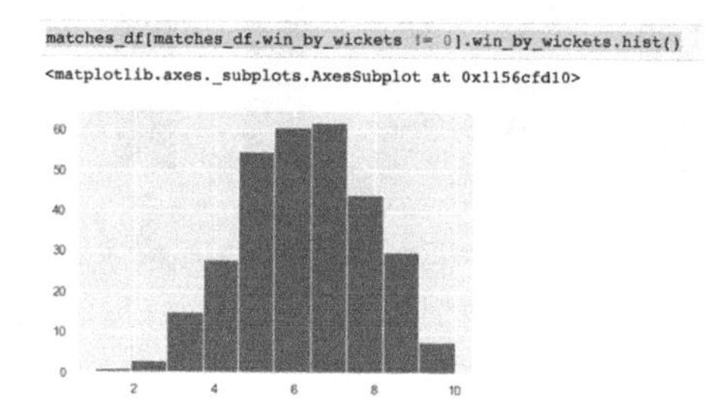

Now we will create box-plots.

```
team_score = deliveries_df.groupby(['match_id', 'batting_
team']).total_runs.sum().reset_index()
top_teams = deliveries_df.groupby('batting_team').total_runs.
sum().nlargest(5).reset_index().batting_team.tolist()
top_teams_df = team_score[team_score.batting_team.isin(top_teams)]
top_teams_df.groupby('batting_team').boxplot(column =
'total_runs', layout=(5,1),figsize=(6,20));
top_teams_df.boxplot( column = 'total_runs',by = 'batting_team',
rot = 20, figsize = (10,4));
```

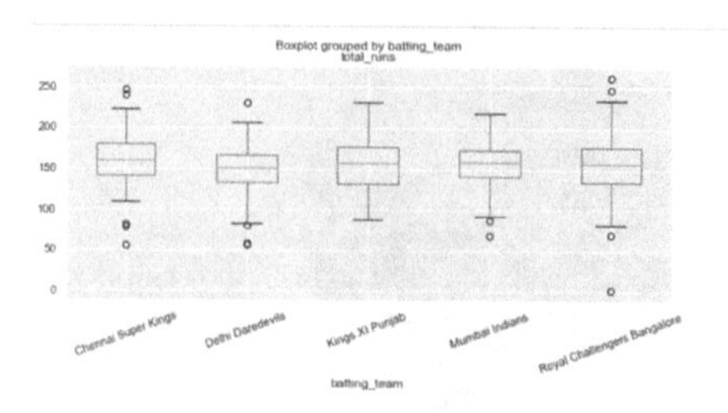

We have developed correlation metrics earlier in the book. Apart from these, there are multiple ways to perform the EDA. Scatter plots, crosstabs, and so on serve as a good tool.

EDA serves as the foundation of a robust ML model. Unfortunately, EDA is often neglected or less time is spent on it. It is a dangerous approach and can lead to challenges in the model at a later stage. EDA allows us to measure the spread of all the variables, understand their trends and patterns, uncover mutual relationships, and give a direction to the analysis. Most of the time, variables which are coming as important during EDA will be found significant by the ML model too. Remember, the time spent on EDA is crucial for the success of the project. It is a useful resource to engage the stakeholders and get their attention, as most of the insights are quite interesting. Visualization makes the insights even more appealing. Moreover, an audience which is *not* from an ML or statistical background will be able to relate to EDA better than to an ML model. Hence, it is prudent to perform a robust EDA of the data.

Now it is time to discuss the ML model-building stage in the next section.

Step 5: ML Model Building

Now we will be discussing the process we follow during the ML model development phase. Once the EDA has been completed, we now have a broad understanding of our dataset. We somewhat are clear on a lot of trends and correlations present in our dataset. And now we have to proceed to the actual model creation phase.

We are aware of the steps we follow during the actual model building. We have solved a number of case studies in the previous chapters. But there are a few points which we should be cognizant of while we proceed with model building. These cater to the common issues we face or the usual mistakes which are made. We will be discussing them in this section.

We will be starting with the training and testing split of the data.

Train/Test Split of Data

We understand that the model is trained on the training data and the accuracy is tested on the testing/validation dataset. The raw data is split into train/test or train/test/validate. There are multiple ratios which are used. We can use 70:30 or 80:20 for train/test splitting. We can use 60:20:20 or 80:10:10 for train/test/validate split, respectively.

While we understand the usage of training and testing data, there are a few precautions which need to be taken:

1. The training data should be representative of the business problem. For example, if we are working on a prepaid customer group for a telecom operator it should be only for prepaid customers and not for postpaid customers, though this would have been checked during the data discovery phase.

2. There should not be any bias while sampling the training or testing datasets. Sometimes, during the creation of training or testing data, a bias is introduced based on time or product. This has to be avoided since the training done will not be correct. The output model will be generating incorrect predictions on the unseen data.

3. The validation dataset should not be exposed to the algorithm during the training phase. If we have divided the data into train, test, and validation datasets, then during the training phase we will check the model's accuracy on the train and test datasets. The validation dataset will be used only once and in the final stage once we want to check the robustness.

4. All the treatments like outliers, null values, and so on should be applied to the training data only and not to the testing dataset or the validation dataset.

5. The target variable defines the problem and holds the key to the solution. For example, a retailer wishes to build a model to predict the customers who are going to churn. There are two aspects to be considered while creating the dataset:

    a. The definition of target variable or rather the label for it will decide the design of the dataset. For example, in Table 5-12 if the target variable is "Churn" then the model will predict the propensity to churn from the system. If it is "Not Churn," it will mean the propensity to stay and not churn from the system.

***Table 5-12.** Definition of Target Variable Changes the Entire Business Problem and Hence the Model*

CustID	Revenue	Gender	Items	Date	Churn
1001	100	M	4	01-Jan-20	0
1002	101	F	5	02-Jan-20	1
1003	102	F	6	04-Jan-20	1
1004	104	F	8	02-Jan-20	0
1001	100	M	4	01-Jan-20	1
1005	105	M	5	05-Jan-20	1

CustID	Revenue	Gender	Items	Date	Not Churn
1001	100	M	4	01-Jan-20	0
1002	101	F	5	02-Jan-20	1
1003	102	F	6	04-Jan-20	1
1004	104	F	8	02-Jan-20	0
1001	100	M	4	01-Jan-20	1
1005	105	M	5	05-Jan-20	1

    b. The duration we are interested in predicting the event. For example, customers who are going to churn in the next 60 days are going to be different from the ones who will churn in 90 days, as shown in Figure 5-6.

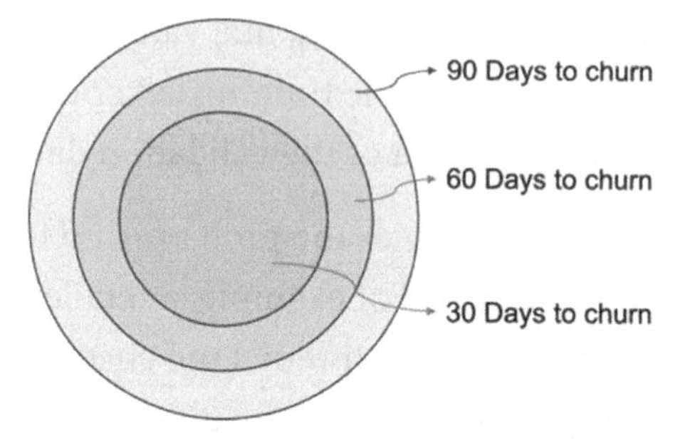

Figure 5-6. *The customer datasets are subsets of each other. For example, customers who are going to churn in 30 days will be a subset of the ones who are going to churn in 90 days.*

Creation of training and testing data is the most crucial step before we do the actual modeling. If there is any kind of bias or incompleteness in the datasets created, then the output ML model will be biased.

Once we have secured the training and testing dataset, we move to the next step of model building.

Model Building and Iterations

Now is the time to build the actual ML model. Depending on the problem at hand and if we are developing a supervised learning solution, we can either choose a regression approach or a classification algorithm. Then we will choose an algorithm from the list of the algorithms available to use. Figure 5-7 shows the relationship between interpretability and accuracy of all the algorithms.

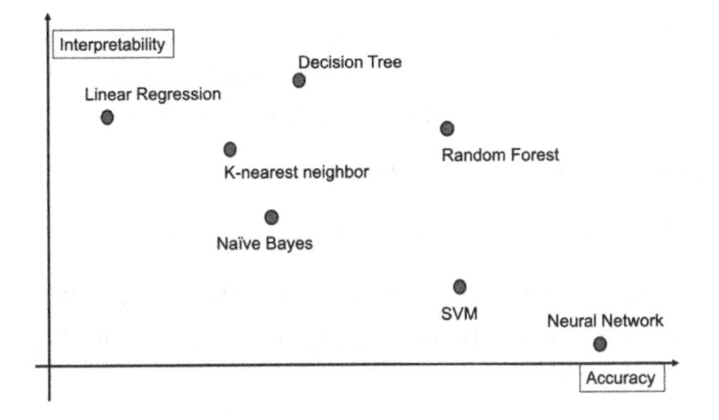

***Figure 5-7.** The arrangement of various algorithms with respect to accuracy and interpretability*

Generally, we start with a base algorithm like linear regression or logistic regression or decision tree. This sets the basic benchmark for us for the other algorithms to break.

Then we proceed to train the algorithm with other classification algorithms. There is a tradeoff which always exists, in terms of accuracy, speed of training, prediction, robustness, and ease of comprehension. It also depends on the nature of the variables; for example, k-nearest neighbor requires numeric variables as it needs to calculate the distance between various data points. One-hot encoding will be required to convert categorical variables to the respective numeric values.

After training data creation, an ML modeling process can be envisioned as the following steps. We are taking an example of a classification problem:

1. Create the base version of the algorithm using logistic regression or decision tree.

2. Measure the performance on a training dataset.

3. Iterate to improve the training performance of the model. During iterations, we try a combination of addition or removal of variables, changing the hyperparameters, and so on.

4. Measure the performance on a testing dataset.

5. Compare the training and testing performance; the model should not be overfitting (more on this in the next section).

6. Test with other algorithms like random forest or SVM and perform the same level of iterations for them.

7. Choose the best model based on the training and testing accuracy.

8. Measure the performance on the validation dataset.

9. Measure the performance on out-of-time dataset.

There is a fantastic cheat sheet provided by sci-kit learn library at `https://scikit-learn.org/stable/tutorial/machine_learning_map/index.html`.

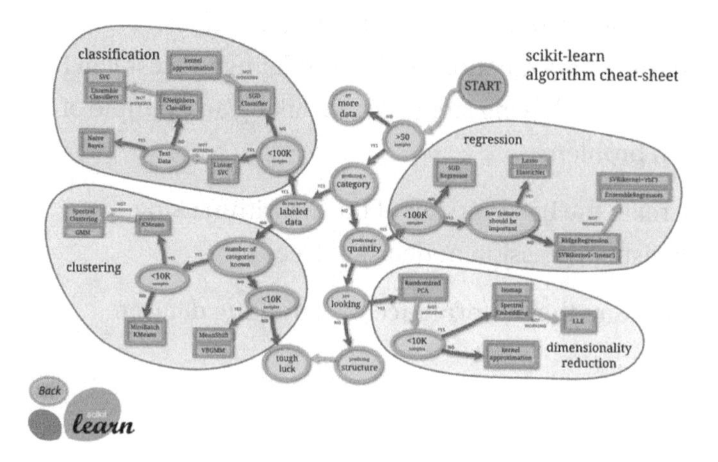

At the end of this step, we would have accuracies of various algorithms with us. We are discussing the best practices for accuracy measurement now.

Accuracy Measurement and Validation

We have discussed the various accuracy measurement parameters for our regression and classification problems. The KPI to be used for accuracy measurement will depend on the business problem at hand. Recall the fraud detection case study discussed earlier: in that case, accuracy is not the correct KPI; instead, recall is the KPI which should be targeted and optimized. Hence, depending on the objective at hand the accuracy parameters will change. During the ML model phase, it is a good practice to compare various algorithms, as shown in Table 5-13.

Table 5-13. *Comparing the Performance Measurement Parameters for All the Algorithms to Choose the Best One*

	Training			Testing			Validation		
	Accuracy	Recall	Precision	Accuracy	Recall	Precision	Accuracy	Recall	Precision
Logistic Regression									
Decision Tree									
Random Forest									
Naïve Bayes									
SVM									

Then we can take a decision to choose the best algorithm. The decision to choose the algorithm will also depend on parameters like time taken to train, time taken to make a prediction, ease of deployment, ease of refresh, and so on.

Along with the accuracy measurement, we also check time taken to prediction and the stability of the model. If the model is intended for real-time prediction, then time taken to predict becomes a crucial KPI.

At the same time, we have to validate the model too. Validation of the model is one of the most crucial steps. It checks how the model is performing on the new and unseen dataset. And that is the main objective of our model: to predict well on new and unseen datasets!

It is always a good practice to test the model's performance on an out-of-time dataset. For example, if the model is trained on a Jan 2017 to Dec 2018 dataset, we should check the performance on out-of-time data like Jan 2019–Apr 2019. This ensures that the model is able to perform on the dataset which is unseen and not a part of the training one. We will now study the impact of threshold on the algorithm and how can we optimize in the next section.

Finding the Best Threshold for Classification Algorithms

There is one more important concept of the most optimum threshold, which is useful for classification algorithms. *Threshold* refers to the probability score above which an observation will belong to a certain class. It plays a significant role in measuring the performance of the model.

For example, consider we are building a prediction model which will predict if a customer will churn or not. Let's assume that the threshold set for the classification is 0.5. Hence, if the model generates a score of 0.5 or above for a customer, the customer is predicted to churn; otherwise not. It means a customer with a probability score of 0.58 will be classified as a "predicted to churn." But if the threshold is changed to 0.6, then the same customer will be classified as a "non-churner." This is the impact of setting the threshold.

The default threshold is set at 0.5, but this is generally unsuitable for unbalanced datasets. Optimizing the threshold is an important task, as it can seriously impact the final results.

We can optimize the threshold using the ROC curve. The best threshold is where True Positive Rate and (1- False Positive Rate) overlap with each other. This approach maximizes the True Positive Rate while minimizing the False Positive Rate. However, it is recommended to test the performance of the model at different values of threshold to determine the best value. The best threshold will also depend on the business problem we are trying to solve.

There is one common problem we encounter during the model development phase: overfitting and underfitting, which we will discussing now.

Overfitting vs. Underfitting Problem

During development of the model, we often face the issue of underfitting or overfitting. They are also referred to as *bias-variance tradeoff*. Bias-variance tradeoff refers to a property of the ML model where while testing the model's performance we find that the model has a lower bias but high variance and vice versa, as shown in Figure 5-8.

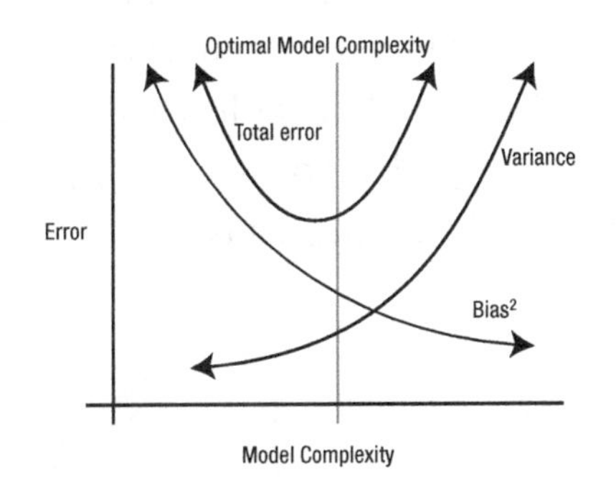

Figure 5-8. *Bias-variance tradoff. Note that the optimal model complexity is the sweet spot where both bias and variance are balanced*

Underfitting of the model is when that model is not able to make accurate predictions. The model is naïve and very simple. Since we have created a very simple model, it is not able to do justice to the dataset as shown in Figure 5-9.

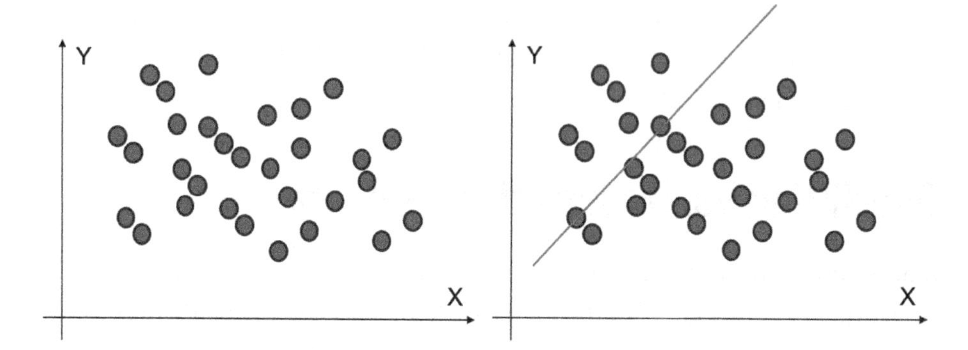

Figure 5-9. *Underfitting where a very simple model has been created*

Overfitting on the other hand is quite the opposite (Figure 5-10). If we have created a model which has good training accuracy but low testing accuracy, it means that the algorithm has learned the parameters of the training dataset very minutely. The model is giving good accuracy for the training dataset but not for the testing dataset.

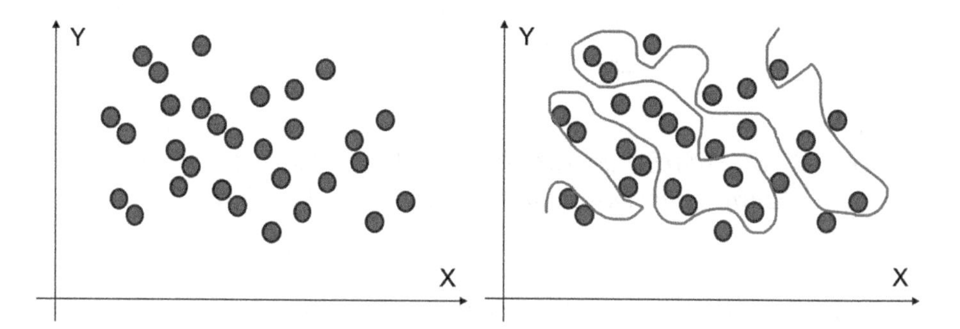

Figure 5-10. *Overfitting where a very complex model has been created*

Underfitting can be tackled by increasing the complexity of the model. Underfitting means that we have chosen a very simplistic solution for our ML approach while the problem at hand requires a higher degree of complexity. We can easily identify a model which is underfitting. The model's training accuracy will not be good and hence we can conclude the model is not even able to perform well on the training data. To tackle underfitting, we can train a deeper tree, or instead of fitting a linear equation, we can fit a nonlinear equation. Using an advanced algorithm like SVM might give improved results.

Tip Sometimes there is no pattern in the data. Even using the most complex ML algorithm will not lead to good results. In such a case, a very strong EDA might help.

For overfitting, we can employ these methods:

1. Maybe the easiest approach is to train with more data. But this might not always work if the new dataset is messy or if the new dataset is similar to the old one without adding additional information.

2. k-fold cross-validation is a powerful method to combat overfitting. It allows us to tweak the respective hyperparameters and the test set will be totally unseen till the final selection has to be made.

   The k-fold validation technique is depicted in Figure 5-11. It is a very simple method to implement and understand. In this technique, we iteratively divide the data into training and testing "folds."

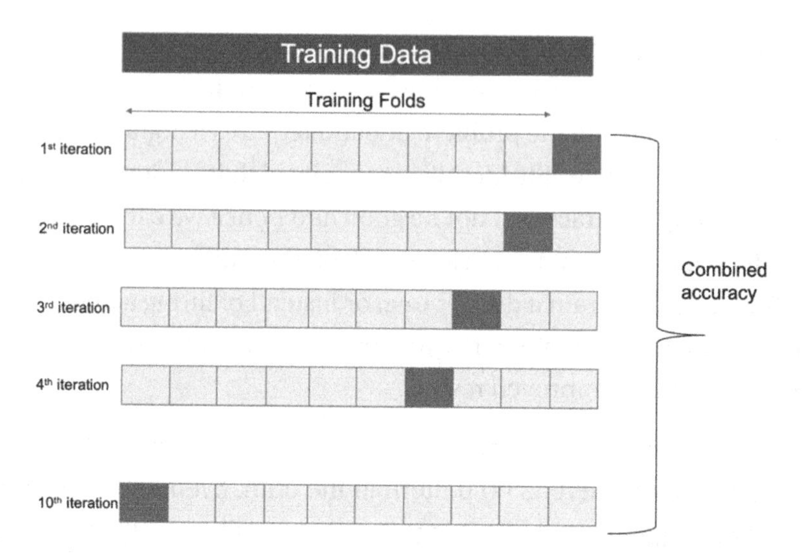

***Figure 5-11.** k-fold cross-validation is a fantastic technique to tackle overfitting*

In k-fold validation, we follow these steps:

1. We shuffle the dataset randomly and then split into k groups.

2. For each of the groups, we take a test set. The remaining serve as a training set.

3. The ML model is fit for each group by training on the training set and testing on the test set (also referred as holdout set).

4. The final accuracy is the summary of all the accuracies.

   It means that each data point gets a chance to be a part of both the testing and the training dataset. Each observation remains a part of a group and stays in the group during the entire cycle of the experiment. Hence, each observation is used for testing once and for training (k–1) times.

The Python implementation of the k-fold validation code is checked in at the Github repo.

3. Ensemble methods like random forest combat overfitting easily as compared to decision trees.

4. Regularization is one of the techniques used to combat overfitting. This technique makes the model coefficients shrink to zero or nearly zero. The idea is that the model will be penalized if a greater number of variables are added to the equation.

   There are two types of regularization methods: Lasso (L1) regression and Ridge (L2) regression. Equation 5-3 may be recalled from earlier chapters.

$$Y_i = \beta_0 + \beta_1 x_1 + \beta_2 x_2 + \ldots + \varepsilon_i \qquad \text{(Equation 5-3)}$$

   At the same time, we have a loss function which has to be optimized to receive the best equation. That is the residual sum of squares (RSS).

   In Ridge regression, a shrinkage quantity is added to the RSS, as shown in Equation 5-4.

   So, the modified $RSS = RSS + \lambda \, \Sigma_{i=1}^{n} \, \beta_i^2$ (Equation 5-4)

Here λ is the tuning parameter which will decide how much we want to penalize the model. If λ is zero, it will have no effect, but when it becomes large, the coefficients will start approaching zero.

In Lasso regression, a shrinkage quantity is added to the RSS similar to Ridge but with a change. Instead of squares, an absolute value is taken, as shown in Equation 5-5.

So the modified $RSS = RSS + \lambda \, \Sigma_{i=1}^{n} |\beta_i|$ (Equation 5-5)

Between the two models, ridge regression impacts the interpretability of the model. It can shrink the coefficients of very important variables to close to zero. But they will never be zero, hence the model will always contain all the variables. Whereas in lasso some parameters can be exactly zero, making this method also help in variable selection and creating sparse models.

5. Decision trees generally overfit. We can combat overfitting in decision tree using these two methods:

    a. We can tune the hyperparameters and set constraints on the growth of the tree: parameters like maximum depth, maximum number of samples required, maximum features for split, maximum number of terminal nodes, and so on.

b. The second method is pruning. Pruning is opposite to building a tree. The basic idea is that a very large tree will be too complex and will not be able to generalize well, which can lead to overfitting. Recall that in Chapter 4 we developed code using pruning in decision trees.

Decision tree follows a greedy approach and hence the decision is taken only on the basis of the current state and not dependent on the future expected states. And such an approach eventually leads to overfitting. Pruning helps us in tackling it. In decision tree pruning, we

i. Make the tree grow to the maximum possible depth.

ii. Then start at the bottom and start removing the terminal nodes which do not provide any benefit as compared to the top.

Note that the goal of the ML solution is to make predictions for the unseen datasets. The model is trained on historical training data so that it can understand the patterns in the training data and then use the intelligence to make predictions for the new and unseen data. Hence, it is imperative we gauge the performance of the model on testing and validation datasets. If a model which is overfitting is chosen, it will fail to work on the new data points, and hence the business reason to create the ML model will be lost.

The model selection is one of the most crucial yet confusing steps. Many times, we are prone to choose a more complex or advanced algorithm as compared to simpler ones. However, if a simpler logistic regression algorithm and advanced SVM are giving similar levels of accuracy, we should defer to the simpler one. Simpler models are easier to comprehend, more flexible to ever-changing business requirements, straightforward to deploy, and painless to maintain and refresh in the future. Once we have made a choice of the final algorithm which we want to deploy in the production environment, it becomes really difficult to replace the chosen one!

The next step is to present the model and insights to the key stakeholders to get their suggestions.

Key Stakeholder Discussion and Iterations

It is a good practice to keep in close touch with the key business stakeholders. Sometimes, there are some iterations to be done after the discussion with the key stakeholders. They can add their inputs using business acumen and understanding.

Presenting the Final Model

The ML model is ready. We have had initial rounds of discussions with the key stakeholders. Now is the time when the model will be presented to the wider audience to take their feedback and comments.

We have to remember that an ML model is to be consumed by multiple functions of the business. Hence, it is imperative that all the functions are aligned on the insights.

Congratulations! You have a functional, validated, tested, and (at least partially) approved model. It is then presented to all the stakeholders and team for a detailed discussion. And if everything goes well, the model is ready to be deployed in production.

Step 6: Deployment of the Model

We have understood all the stages in model development so far. The ML model is trained on historical data. It has generated a series of insights and a compiled version of the model. It has to be used to make predictions on the new and unseen dataset. This model is ready to be put into the production environment, which is referred to as *deployment*. We will examine all the concepts in detail now.

ML is a different type of software as compared to traditional software engineering. For an ML model, several forces join hands to make it happen. It is imperative we discuss them now because they are the building blocks of a robust and useful solution.

Key elements are shown in Figure 5-12.

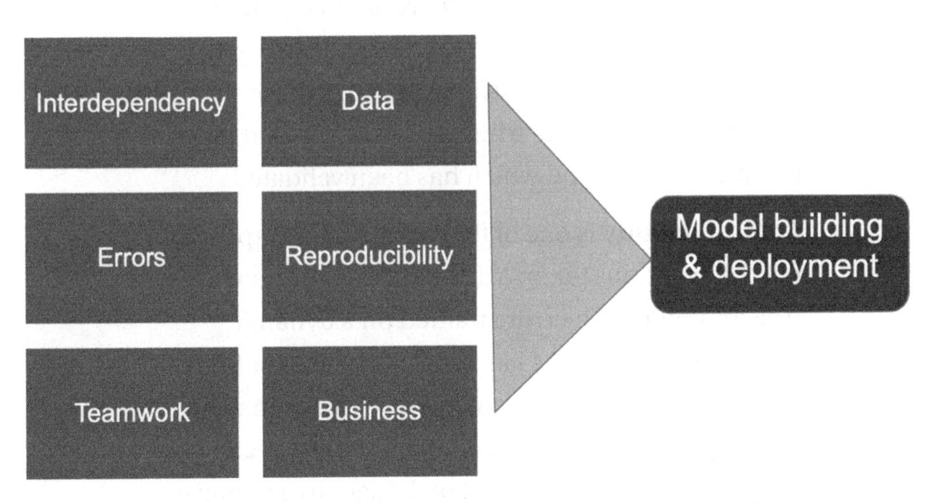

Figure 5-12. *Key elements of an ML solution which are the foundation of the solution*

1. **Interdependency** in an ML solution refers to the relationship between the various variables. If the values of one of the variables change, it does make

an impact on the values of other variables. Hence, the variables are interdependent on each other. At the same time, the ML model sometimes is quite "touchy" to even a small change in the values of the data points.

2. **Data** lies at the heart of the model. Data powers the model; data is the raw material and is the source of everything. Some data points are however very unstable and are always in flux. It is hence imperative that we track such data changes while we are designing the system. Model and data work in tandem with each other.

3. **Errors** happen all the time. There are many things we plan for a deployment. We ensure that we are deploying the correct version of the algorithm. Generally, it is the one which has been trained on the latest dataset and which has been validated.

4. **Reproducibility** is one of the required and expected but difficult traits for an ML model. It is particularly true for models which are trained on a dynamic dataset or for domains like medical devices or banks which are bound with protocols and rules. Imagine an image classification model is built to segregate good vs. bad products based on images in a medical devices industry. In a case like that, the organization will have to adhere to protocols and will be submitting a technical report to the regulatory authorities.

5. **Teamwork** lies at the heart of the entire project. An ML model requires time and attention from a data engineer, data analyst, data scientist, functional consultant, business consultant, stakeholder, dev ops engineer, and so on. For the deployment too, we need guidance and support from some of these functions.

6. **Business** stakeholder and sponsors are the key to a successful ML solution. They will be guiding how the solution can be consumed and will be used to resolve the business problem.

Tip Deployment is not an easy task; it requires time, teamwork, and multiple skills to handle.

Before we start the deployment of the ML model, we should be clear on a few questions. Apart from the key elements which we discussed in the last paragraphs, we should also be cognizant of the following points about our ML model:

1. We have to check if the ML model is going to work in real time or in a batch prediction mode. For example, checking if an incoming transaction is fraudulent or not is an example of a real-time check. If we are predicting if a customer will default or not, that's not a real-time system but a batch prediction mode and we have the luxury to make a prediction maybe even the next day.

2. The size of the data we are going to deal with is a big question to be answered. And so it is important to know the size of the incoming requests. For

example, if an online transaction system expects 10 transactions per second as compared to a second system which expects 1000 transactions, load management and speed of decision will make a lot of difference.

3. We also make sure to check the performance of the model in terms of time it takes to predict, stability, what the memory requirements are, and how is it producing the final outputs.

4. Then we also decide if the model requires on-off training or real-time training.

   a. On-off training is where we have trained the model of historical data and we can put it in production till its performance decays. Then we will be refreshing the model.

   b. Real-time training allows the ML model to be trained on a real-time dataset and then get retrained. It allows the ML model to always cater to the new dataset.

5. The amount of automation we expect in the ML model is to be decided. This will help us to plan the manual steps better.

Once we have answered these questions, we are clear on the purpose and business logic of the model. After a thorough round of testing, the model is ready to be moved to the production environment. We should then work on these points:

1. It all starts with having a database. The database can be in the cloud or a traditional server. It contains all the tables and sources from which the model is going to consume the incoming data as shown in Figure 5-13. Sometimes, a separate framework feeds the data to the model.

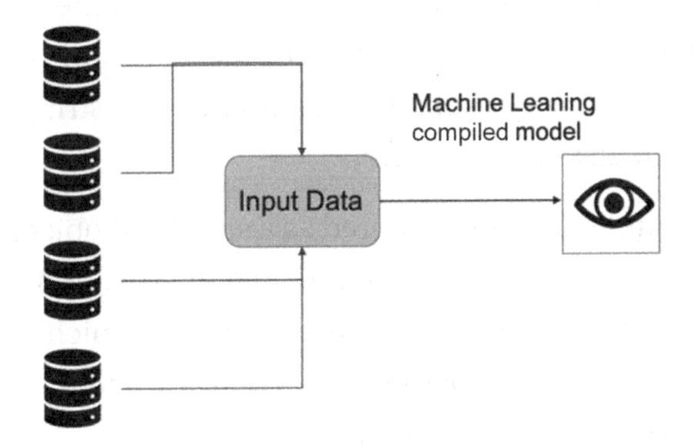

Figure 5-13. *An ML model can take input data from multiple sources*

2. In case a variable is not a raw variable but a derived one, the same should have been taken care of in the creation of the data source for the data. For example, consider for a retailer, the incoming raw data is customer transactions details of the day. We built a model to predict the customer churn, and one of the significant variables is the customer's average transaction value (total revenue/number of transactions); then this variable will be a part of the data provider tables.

3. This incoming data is unseen and new to the ML model. The data contains the data in the format expected by the compiled model. If the data is not

in the format expected, then there will be a runtime error. Hence, it is imperative that the input data is thoroughly checked. In case we have an image classification model using a neural network, we should check that the dimensions of the input image are the same as expected by the network. Similarly, for structured data it should be checked that the input data is as expected and is checked including the names and type of the variables (integer, string, Boolean, data frame, etc.).

4. The compiled model is stored as a serialized object like .pkl for Python, .R file for R, .mat for MATLAB, .h5, and so on. This is the compiled object which is used to make the predictions. There are a few formats which can be used:

    a. Pickle format is used for Python. The object is a streamed object which can be loaded and shared across. It can be used at a later point too.

    b. ONNX (Open Neural Network Exchange Format) allows storing of predictive models. Similarly, PMML (Predictive Model Markup languages) is another format which is used.

5. The objective of the model, the nature of predictions to be made, the volume of the incoming data, real-time prediction or on-off prediction, and so on, allow us to design a good architecture for our problem.

6. When it comes to the deployment of the model, there are multiple options available to us. We can deploy the model as follows:

a. **Web service**: ML models can be deployed as a web service. It is a good approach since multiple interfaces like web, mobile, and desktop are handled. We can set up an API wrapper and deploy the model as a web service. The API wrapper is responsible for the prediction on the unseen data.

   Using a web service approach, the web service can take an input data and convert into a dataset or data frame as expected by the model. Then, it can make a prediction using the dataset (either continuous variable or a classification variable) and return the respective value.

   Consider this. A retailer sells specialized products online. A lot of the business comes from repeat customers. Now a new product is launched by the retailer. And the business wants to target the existing customer base.

   So, we want to deploy an ML model which predicts if the customer will buy a newly launched product based on the last transactions by the customer. We have depicted the entire process in Figure 5-14.

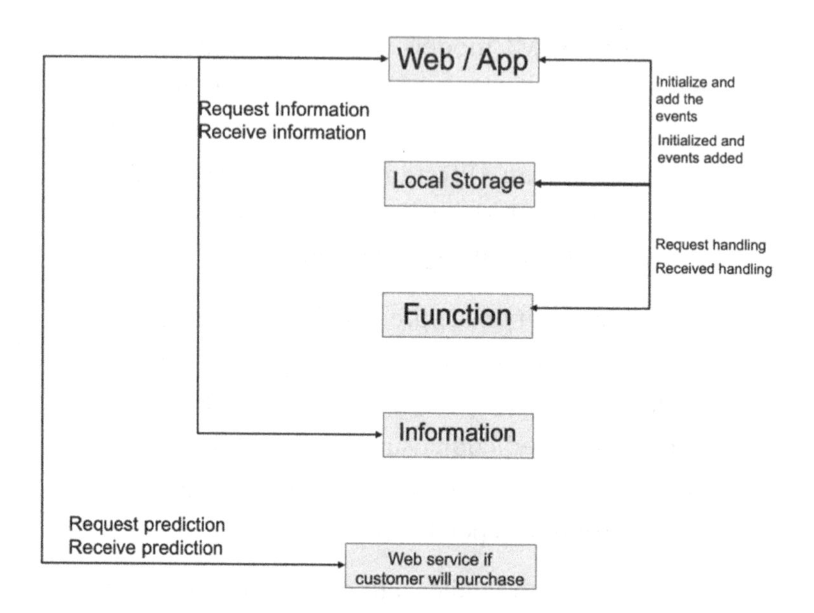

Figure 5-14. *Web service–based deployment process*

We should note that we already have historical information on the customer's past behavior. Since there is a need to make a prediction on a new set of information, the information has to be merged and then shared with the prediction service.

As a first step, the website or the app initializes and makes a request to the information module. The information module consists of information about customer's historical transactions, and so on. It returns this information about the customer to the web application. The web app initializes the customer's profile locally and stores this information locally. Similarly, the new chain of events or the trigger which is happening at the web or in the app is also obtained. These data points (old customer details and new data values) are shared with a function or a wrapper function which updates the information received based on the new data points. This updated information is then shared with the prediction web service to generate the prediction and receive the values back.

In this approach, we can use AWS Lambda functions, Google Cloud functions, or Microsoft Azure functions. Or we can use containers like Docker and use it to deploy a flask or Django application.

b. **Integrated with the database**: Perhaps this approach is generally used. As shown in Figure 5-15, this approach is easier than a web service–based one but is possible for smaller databases.

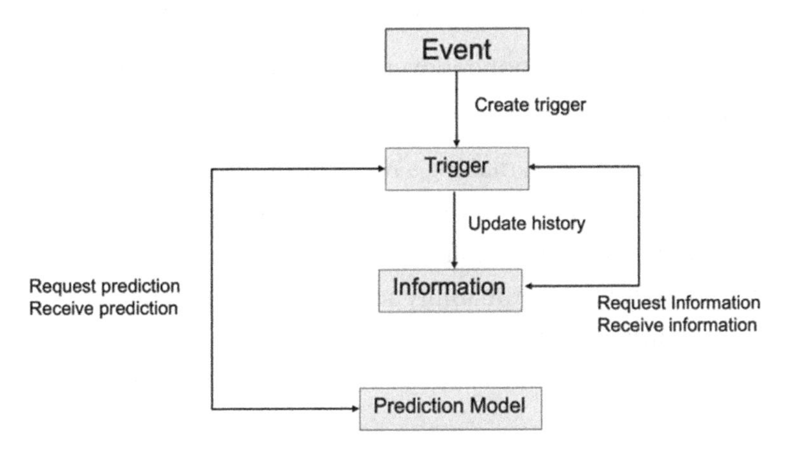

***Figure 5-15.** Integration with database to deploy the model*

In this approach, the moment a new transaction takes place an event is generated. This event creates a trigger. This trigger shares the information to the customer table where the data is updated, and the updated information is shared with the event trigger. Then the prediction model is run, and a prediction score is generated to predict if the customer will make a purchase of the newly launched product. And finally, the customer data is again updated based on the prediction made.

This is a simpler approach. We can use databases like MSSQL, PostGres, and so on.

We can also deploy the model into a native app or as a web application. The predictive ML model can run externally as a local service. It is not heavily used. Once the model is deployed, the predictions have to be consumed by the system.

7. The next step to ponder over is consumptions of the predictions done by the ML model.

    a. If the model is making a *real-time* prediction as shown in Figure 5-16, then the model can return a signal like pass/fail or yes/no. The model can also return a probability score, which can then be handled by the incoming request.

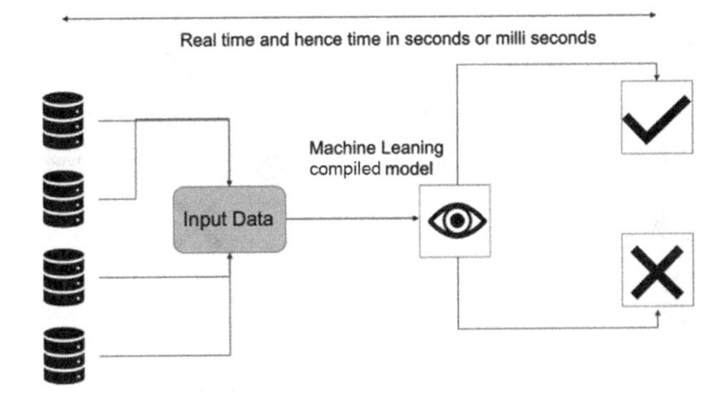

Figure 5-16. *Real-time prediction system takes seconds or milliseconds to make a prediction*

b. If the model is not a real-time model, then it is possible that the model is generating probability scores which are to be saved into a separate table in the database. It is referred to as *batch prediction*, as shown in Figure 5-17. Hence, the model-generated scores have to be written back to the database. It is not necessary to write back the predictions. We can have the predictions written in a .csv or .txt file, which can then be uploaded to the database. Generally, the predictions are accompanied by all the details of the raw incoming data for which the respective prediction has been made.

The predictions made can be continuous in case of a regression problem or can be a probability score/class for a classification problem. Depending on the predictions, we can make configurations accordingly.

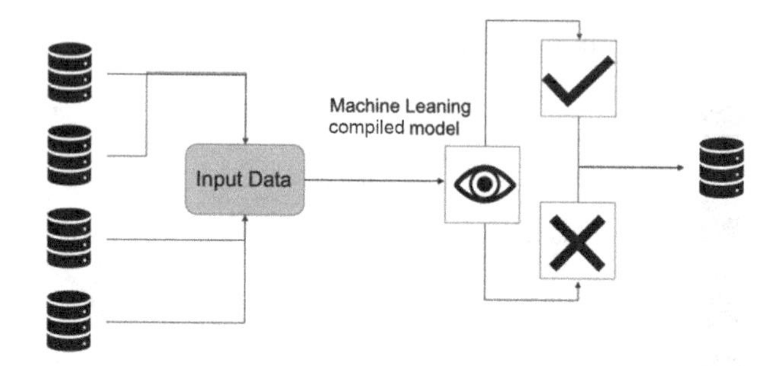

Figure 5-17. *Batch prediction is saved to a database*

8. We can use any cloud platform which allows us to create, train, and deploy the model too. Some of the most common options available are

    a. AWS SageMaker

    b. Google Cloud AI platform

    c. Azure Machine Learning Studio, which is a part of Azure Machine Learning service

    d. IBM Watson Studio

    e. Salesforce Einstein

    f. MATLAB offered by MathWorks

    g. RapidMiner offers RapidMiner AI Cloud

    h. TensorFlow Research Cloud

Congratulations! The model is deployed in the production and is making predictions on real and unseen dataset. Exciting, right?

Step 7: Documentation

Our model is deployed now we have to ensure that all the codes are cleaned, checked in, and properly documented. Github can be used for version control.

Documentation of the model is also dependent on the organization and domain of use. In regulated industries it is quite imperative that all the details are properly documented.

Step 8: Model Refresh and Maintenance

We have understood all the stages in model development and deployment now. But once a model is put into production, it needs constant monitoring. We have to ensure that the model is performing always at a desired level of accuracy measurements. To achieve this, it is advised to have a dashboard or a monitoring system to gauge the performance of the model regularly. In case of non-availability of such a system, a monthly or quarterly checkup of the model can be done.

Model refresh can follow the approach shown in Figure 5-18. Here we have shown the complete step-by-step approach.

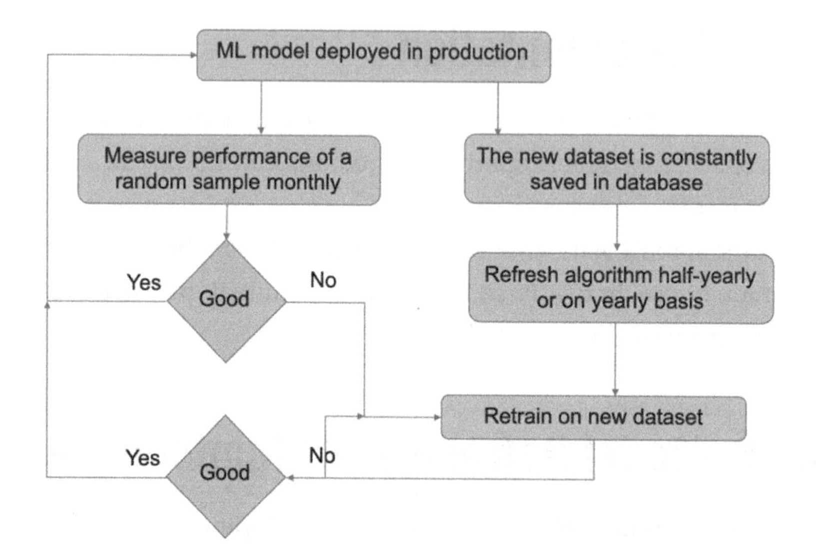

Figure 5-18. *ML model refresh and maintenance process*

Once the model is deployed, we can do a monthly random check of the model. If the performance is not good, the model requires refresh. It is imperative that even though the model might not be deteriorating, it is still a good practice to refresh the model on the new data points which are constantly created and saved.

With this, we have completed all the steps to design an ML system, how to develop it from scratch, and how to deploy and maintain it. It is a long process which is quite tedious and requires teamwork.

With this, we are coming to the end of the last chapter and the book.

Summary

In this last chapter we studied a model's life—from scratch to maintenance. There can be other methods and processes too, depending on the business domain and the use case at hand.

With this, we are coming to the end of the book. We started this book with an introduction to ML and its broad categories. In the next chapter, we discussed regression problems, and then we solved classification problems. This was followed by much more advanced algorithms like SVM and neural networks. In this final chapter, we culminated the knowledge and created the end-to-end model development process. The codes in Python and the case studies complemented the understanding.

Data is the new oil, new electricity, new power, and new currency. The field is rapidly increasing and making its impact felt across the globe. With such a rapid enhancement, the sector has opened up new job opportunities and created new job titles. Data analysts, data engineers, data scientists, visualization experts, ML engineers—they did not exist a decade back. Now, there is a huge demand of such professionals. But there is dearth of professionals who fulfill the rigorous criteria for these job descriptions. The need of the hour is to have *data artists* who can marry business objectives with analytical problems and envision solutions to solve the dynamic business problems.

Data is impacting all the business, operations, decisions, and policies. More and more sophisticated systems are being created everyday. Data and its power are immense: we can see examples of self-driving cars, chatbots, fraud detection systems, facial recognition solutions, object detection solutions, and so on. We are able to generate rapid insights, scale the solutions, visualize, and take decisions on the fly. The medical industry can create better drugs, the banking and financial sectors can mitigate risks, telecom operators can offer better and more stable network coverages, and the retail sector can provide better prices and serve the customers better. The use cases are immense and still being explored.

But the onus lies on us on how to harness this power of data. We can put it to use for the benefit of mankind or for destroying it. We can use ML and AI for spreading love or hatred—it is our choice. And like the cliché goes—with great power comes great responsibility!

EXERCISE QUESTIONS

Question 1: What is overfitting and how can we tackle it?

Question 2: What are the various options available to deploy the model?

Question 3: What are the types of variable transformations available?

Question 4: What is the difference between L1 and L2 regularization?

Question 5: Solve the exercises we did in previous chapters and compare the accuracies after treating missing values and correlated variables.

Question 6: What are the various steps in EDA?

Question 7: Download the NFL dataset from the following link and treat for the missing values: `https://www.kaggle.com/maxhorowitz/nflplaybyplay2009to2016`.

Question 8: How can we define an effective business problem?

Index

A

B

C

V. Verdhan, *Supervised Learning with Python*,
https://doi.org/10.1007/978-1-4842-6156-9

D

E, F

G

H, I

J

K

L

M

S

T

U

V

W, X, Y, Z